Annual Meeting of the Society for Machinery Failure Prevention Technology (MFPT 2020)

Savannah, Georgia, USA
4 - 7 August 2020

ISBN: 978-1-7138-2471-8

Printed from e-media with permission by:

Curran Associates, Inc.
57 Morehouse Lane
Red Hook, NY 12571

Some format issues inherent in the e-media version may also appear in this print version.

Printed by Curran Associates, Inc. (2021)

For permission requests, please contact Society for Machinery Failure Prevention Technology
at the address below.

MFPT Society
715 Twining Road
Suite 135
Dresher PA 19025

Phone: (215) 346-2870
Fax: (215) 352-6902

https://mfpt.org

Additional copies of this publication are available from:

Curran Associates, Inc.
57 Morehouse Lane
Red Hook, NY 12571 USA
Phone: 845-758-0400
Fax: 845-758-2633
Email: curran@proceedings.com
Web: www.proceedings.com

FNAME	LNAME	Title	Abstract
Eric	Bechhoefer	Signal Processing to Reduce the Effect of Gear Dynamics	An improved conditioning monitoring technique is provided for rotating components in gearboxes that accounts for gear system dynamics. A rotation rate for the component is generated from vibration data by estimating the rotation rate based on a tachometer measurement of another shaft and the shaft ratio. This estimated rotation rate is used, together with known configuration of the component, to estimate a known gear mesh frequency of the component. By filtering for a range of frequencies around the gear mesh frequency based on variation in the shaft rate, the gear mesh frequency can be determined and from that signal, an actual rotation rate for the component can be determined. The actual rate can then be used in deriving an analytic vibration spectrum for the component that is not degraded due to gear system dynamics effects. This paper goes into analysis technique needed for a guided reconstruction of the tachometer from measured vibration in order to correct for geared shaft dynamics. We cover some examples, and present results based on real world helicopter gearbox data. This technique is computationally intensive, but due to newly available microcontrollers, should be implemented early next year in an edge processing, smart sensor. This can greatly reduce the cost of system implementation. For example, it is anticipated that this technique could be used on the compressor section of of the PT6T3 on a Bell 212, which would normally require two additional tachometer interfaces PG. 1
Eric	Bechhoefer	Improved Spectral Estimation of Signals using Quadratic Interpolation	Common in many fields of science and industry is the estimation of the magnitude and frequency of signals generated by periodic, sinusoidal signals. This process is ubiquitous and the foundation of many standards and procedures, and often involves use of the Fourier transform. In application, this is done with a Fast Fourier Transform (FFT). The FFT is defined only for frequencies that are exactly integers of the sample rate (fs). For other frequencies, this causes spectral leakage and error in the estimation of the FFT amplitude for frequency. On average, the magnitude error is 4%. This paper discuses an improved technique for the estimation of magnitude and frequency of a sinusoidal signal using the FFT, which reduces the magnitude error to a fraction of a percent. The improved estimation uses a transformation of the data to make the spectral peak parabolic, from which a quadratic solver is used to estimate the true magnitude peak value and frequency. This method can be proved to be optimal by reducing the FFT error in sum of square errors sense. PG. 13
Mantosh	Bhattacharya	Right sizing of gear box for a centrifugal compressor with synchronous motor as driver	In oil & gas industry, for high speed centrifugal compressors speed increasing gear boxes are used to achieve high polytrophic head and flow. API standards and project specifications (as addendum to API) are the guidelines used to design compressor trains. The proposed paper presents a study in order to stimulate a thought process for end users / purchasers to prudently use API 613, 5th edition standard, where gear box sizing can be hugely impacted by driver motor rating, in order to comply to the guideline for mechanical rating of gear boxes bulleted clause stated as "For electric motor drivers, the gear unit rated power shall be the motor nameplate rating multiplied by the motor service factor". As the gear box size is increased to meet above requirement, torsional parametric changes result in a larger coupling to avoid crucial torsional vibration modes affecting the driven part particularly during start up. Various options to counter torsional instability in driven part may be proposed by OEM but they may not be explicitly categorized in API 671 as a dry flexible coupling. The larger coupling may lead to higher windage loss, causing very high temperature in coupling guard violating API 671 requirement. The issue becomes more aggravated due to a synchronous motor as driver with a hot start requirement, which may lead to redesign of the complete motor to avoid overheating during start up. The ramifications to incorporate this requirement are an increase in overall foot print of the train, a new and bigger motor, with larger gear, lubrication oil requirements and oil cooler size. A detailed internal study, including sensitivity analysis of subject gear box, for varied operating process conditions reveals that incorporation of such a clause into the specification may not be very fruitful, as such a higher polytrophic head requires a higher peripheral speed where a fixed speed driver is used. Additionally, a higher flow to extract the maximum installed driver power may lead to an operating point at the stonewall zone, based on steep slope of performance curve of subject compressor. The paper finally concludes that this particular API clause should be prudently applied on specifications based on well production and depletion profile with a future plan of bundle replacement of the compressor. PG. 24
Mantosh	Bhattacharya	Cases of vibrations in high speed pinion in low load condition in API 613 turbo-gears.	In Oil & Gas, Petrochemical and Refinery industries, API 613 compliant double helical speed increasing gear boxes are widely used for high speed turbo-compressor trains. During operation of such trains, the selected gear box model normally has vibrations within acceptable limits as mandated by API standard or jointly agreed by OEM (Original equipment manufacturer) and End user. However, during low load operation during mechanical run test / coast down of a turbomachinery train the pinion of speed increasing gear box may show up high vibration and cause concerns. Gear box vibration signature also gives a tell tale indication of compressor flow related issue as well . The proposed paper discusses several case studies where pinion vibration issues faced during various scenario such as API 613 no / low load full speed run / during coast down or operating the train below actual performance map at site. The case studies include high vibration at pinion non- drive end, low frequency vibration at bull gear and pinion, vibration due to casing resonance, superficial indication of pinion whirl & whip ,vibration caused by torsional lateral interaction during coast down of turbomachine and sometimes due to flow related issue . The paper concludes that correcting the such vibration issues at vendor works / site during no load – low load condition based on test results does not necessarily mean a silver bullet solution and may not guarantee the smooth operation of gear box at rated load, hence detailed analysis is needed. A deeper understanding is needed to address such vibration phenomena as case to case basis to decide whether issue to be considered as genuine or benign and further studies and actions are required . PG. 40
Nazli	EshKalak	A Two-Plane Balancing Method for Detection and Correction of Shaft Unbalance	In rotating machinery, vibrations generated due to unbalance are considered to be a significant problem that adversely affects the performance of these machine systems. Hence, monitoring and minimizing such vibrations is of importance in order to prevent consequences such as unexpected damage and catastrophic failures. This study presents numerical and experimental investigations of two-plane balancing to suppress the vibration of a rigid rotating shaft by means of an influence coefficient method. The main goal is to gain solid understanding of the balancing hardware and algorithm. In order to examine the influence coefficient method, a test stand containing a rotating shaft supported by two ball bearings is provided. Balancing is attained by adding a set of weights into two circular balancing planes that are attached to the bearings. Experiments are conducted by utilizing the sensing and data acquisition hardware that consists of two accelerometers (one per bearing), a mag-pickup (MPU), and a data acquisition system. Each balancing iteration is composed of a three-step process. First, the initial vibrations of the two balancing planes, together with the resultant vibrations after adding a trial weight onto the circumference of each plane, are measured. Then, the amplitude and phase information is extracted from the three vibration measurements, and the influence coefficient matrix, which represents the correction plane interferences between two planes, is calculated. It is worth mentioning that the correction plane interference is a cross effect that occurs when an indicated vibration at one balancing plane is a result of an existing unbalance on the other balancing plane. Finally, the correction weights required to offset the initial vibrations on the two balancing planes are obtained. Given the practical constraints on the number and locations of balancing holes and the types of trial/correction weights on each balancing plane, a calculated correction weight may not be directly used for balancing purposes, and in such cases, weight splitting by employing the least squares method is conducted to approximate the calculated correction weight. The approximate correction weights obtained from weight splitting are then added to the balancing plane in order to minimize the residual vibration of the shaft. PG. 56

J	Pattabiraman	Management of Stress-- A Mechanical System Simulation Approach	With many years of interaction with great many institutions and Industries and meeting with people of all backgrounds in India and abroad, one tends to become more philosophical in nature. This study is prompted by the lack of peace and harmony in the human lives leading to global unrest, poverty and upsurge of emotions, creating disturbing thoughts and events all over. This topic assumes even more importance in view of the Pandemic Covid-19 which is posing a threat to safe and peaceful living in local and Global dimensions. During the Covid times, people are facing tremendous stress due to lockdown curbing social meetings, professional/business meetings in person, gatherings, and travel and forced to remain indoors for attending online classes, office work, shopping or entertainment. If professionals divert their attention on the human aspects as well, apart from their engineering and scientific inputs to society, their job would be much more rewarding and would have a direct uplifting effort on the society and thus create more satisfaction for them. In the management sciences, especially for Human Resources Development personnel, this would be essential. A mechanical engineer is likely to deal with several types of systems and mathematically simulate them according to the laws of physics and nature, with a view to understanding their behaviour better. My endeavour through this paper is to share ideas that provide some means to simulate human stress behaviour through an understanding of mechanical dynamic systems and their constituent elements, thus establishing a correspondence between the two systems. The article does not contain any equations or experimental findings but merely conveys the similarities between the two by an appropriate simulation model and gives scope for improving the behavioral response especially under stress in the fast-moving world of today. I felt it necessary to communicate to the scientific and engineering community, the parallelism in the approach and the apparent equivalence between the two analogous systems. Human behaviour plays a much more significant role especially in today's context, to understand, appreciate and utilize the fruits of scientific advancements and Inventions, without which the influence of science on society may not assure the desired comfort and safety in life.The right understanding and remaining calm even under stress will definitely help in the correct decision making. Identification of the real owner and reducing the dominant role of 'I ' and 'Mine ' and increasing the role of ' We ' and ' Ours ' will help maintain inner peace and harmony with the mind getting free from anxiety and worry. PG. 68
Marc	Pepi	ADDITIVELY MANUFACTURED METAL POWDER GAS ATOMIZED IN A MOBILE FOUNDRY FROM RECYCLED SCRAP FOR LIGHTWEIGHT PROTECTION APPLICATIONS	US Army Directive 2019-29 ("Enabling Readiness and Modernization Through Advanced Manufacturing") envisions advanced manufacturing transforming battlefield logistics through on-demand fabrication of parts close to the point of need. Towards this end goal, the Combat Capabilities Development Center – Army Research Laboratory (CCDC-ARL) is researching technologies that will make it possible for the warfighter to manufacture or repair components at the point of need with scrap materials on-hand, thus increasing operational readiness while reducing the military logistics footprint. The objective of this project was to show a proof-of-concept of additively manufacturing metal powder that was gas atomized from scrap rolled homogeneous armor (RHA) steel in a Mobile Foundry housed within a shipping container. The resultant alloy powder was subjected to laser powder bed fusion (LPBF), for builds that were tested metallurgically and ballistically. Cursory results showed that the builds from powder made from recycled RHA had a martensitic structure, exhibited excellent mechanical properties, and improved ballistic properties versus this alloy in the wrought condition. PG. 75
Zlatan	Racic	Resonance Effect, Critical and Resonance Velocities Applied to Diagnostics, Stability and Balancing Methods of Turbine and Generator Rotors over 40 MVA	Standard power generation industry practices of inspection, machining, balancing, assembly and bearings and couplings alignment have been developed and streamlined by OEMs for newly manufactured rotors, and implicitly assume that rotors being installed are within design dimensional specifications and tolerances. However, applying these standard practices in the service industry can fall short of expected performance, when dealing with continuous rotors with inherent eccentricities outside such specifications. In practice the measured rotors' runouts are equated "unbalances", which excite specific mode at specific operating angular velocity. In this paper we will determine the range of velocities of unsteady rotor's rotation. It is shown that resonance velocities of continuous rotor are not dependent on rotor natural fundamental harmonic resonance frequency, but on the value of accelerating rotating moment that ensures increase of rotor's angular velocity and transfer of rotating and non-rotating reference frames through critical velocity. It is explained also that the critical velocity is not equivalent to harmonic resonance oscillation theory in physics. The specific role of rotor inherent resonance in preparation of rotor to transfer rotating and inertial reference frames through critical velocity range is addressed. It is shown that the resonance phenomenon changes the location of center of mass of the rotor body relative to the journals rotation axis and rotor's geometric center as well as creates conditions for self centering of rotor's body mass axis. The presented research is based on inertial theory of rotor's dynamics based on equation of motion developed impulse and momentum method. The result of this research and understanding of continuous rotor dynamics is of particular importance when developing balancing methods of rotors in power generation industry of T-G set over ~40 MVA of generating electric power, which should be introduced as a separate subsection of the ISO21940-1 standard. PG. 89
Jeremy	Sheldon	Combining Wear Debris and Vibration for a More Complete Understanding of Machinery Health	This paper thoroughly discusses vibration and oil debris monitoring. The application of each sensing technology to an actual real-world wind turbine application will be presented in detail. A focus of the paper will be the use of vibration and oil debris together to improve overall CBM metrics. By combining oil debris and vibration, the CBM practitioner will realize improved visibility to the health of their target machine. PG. 110

Signal Processing to Improve Gear Fault Diagnostics in the presence of Gear Dynamics

Eric Bechhoefer[1], Yalin Ozturk[2]

[1] *GPMS Inc, Cornwall, VT, 05753, USA*
[2] *Turkish Aerospace, Ankara, Turkey*

Abstract: An improved conditioning monitoring technique is provided for rotating components in gearboxes that account for gear system dynamics. A rotation rate for the component is generated from vibration data by estimating the rotation rate based on a tachometer measurement of another shaft and the shaft ratio. This estimated rotation rate is used, together with the known configuration of the component, to estimate a known gear mesh frequency of the component. By filtering for a range of frequencies around the gear mesh frequency based on variation in the shaft rate, the gear mesh frequency can be determined, and from that signal, an actual rotation rate for the component can be determined. The exact rate can then be used in deriving an analytic vibration spectrum for the component that is not degraded due to gear system dynamics effects. Further, this allows the calculation of the gearbox transfer function.

Keywords: Gear Dynamics, TSA, Analytic Signal, Fourier Transform

Introduction: Condition monitoring of components in machines is a technique in which a measure of the health of components can be realized by using sensors. Condition monitoring improves asset reliability and allows for improved availability by allowing for opportunistic maintenance. Also, condition monitoring has been successful in improving safety. For gearboxes, condition monitoring typically involves the use of accelerometers to measure vibrational energies associated with rotating components. The condition monitoring system performs analysis of the accelerometer data to extract features that are indicative of component damage and fatigue. Vibration health monitoring systems may monitor different aspects of main gearbox components, such as gears, shafts, and bearings, including shaft order 1 (first harmonic of the shaft under analysis), shaft order 2, gear mesh frequencies, gear tooth indicators, bearing wear indicators, and modulation of web tone epicyclic gear indictors.

The monitoring of gears and shafts is typically based on analyses that use the Time Synchronous Average (TSA) of measured vibrational signals (Ref 1). Time synchronous averaging is a method for reducing noise in spectra of complex signals as well as for extracting more information from the signals. The TSA may be used in many analyses, including Residual/Different analysis, the Energy Operator, the Narrow Band Analysis, the Amplitude Modulation Analysis, the Frequency Modulation Analysis, and other analyses (Ref 2, 3).

Limitations of Gear Analysis: It was recently observed that the measured rotation rate might not result in a precise rotation rate of the component of interest if the component of interest is displaced physically apart (in terms of gear dynamics) from the shaft that the tachometer is measuring. In other words, the actual rotation rate can be affected by the phase error. This variation in actual versus measured rotational rates is due to the global effect of gear system dynamics. It is deleterious to an analysis of a time-synchronous averaged vibration spectrum used for condition monitoring (e.g., a similar phenomenology as jitter on the tachometer signal). That is, if the tachometer is not mounted near the component (shaft or gear) under analysis, the analysis is degraded due to phase errors introduced by the torsional dynamics of the system. The extent of this degradation depends on the dynamic characteristics of the geared shaft system and the operational speed, i.e., the margin between the operational speed and related critical speeds (natural frequencies) in terms of torsional dynamics.

An improved conditioning monitoring technique is presented that accounts for gear system dynamics by determining the actual rotation rate for a component of interest from vibration data. The determined rotation rate is, in this way, corrected for gear system dynamics and allows for improved signal analysis of the component of interest so that faults or defects can be detected earlier and more accurately. The analytic signal is developed using an ideal filter in a single functional process, and noise is removed. This results in an improved TSA spectrum for rotating components in gearboxes, and TSA spectra are fundamental to gear fault diagnostics using vibration-based analysis.

Gear Analysis Based on the Time Synchronous Average: The time synchronous average (TSA) signal is generated by resampling data to account for variation in shaft rates, which confers many benefits. An assumption of the Fourier transform is stationarity, i.e., that the signal does not change over time. Due to limits in the bandwidth of the motor/engine controllers, there is always some change in the shaft rate. This change in shaft rate can be measured directly with a tachometer, which is apparent from the data shown in **Error! Reference source not found.**, which shows the power turbine shaft rate of a light helicopter over time. The tachometer sample rate is 48,000,000 samples per second and accurately measures the speed of the engine power turbine shaft, and, as such, when an appropriate ratio is applied, it can be used to find the shaft rates of other shafts in the gearbox. Note that the shaft rate is not constant. The variation is due to changes in torque coming from the main, tail rotor, and the control cycle of the FADEC. Any inputs from the flight control and in forwarded flight, the added lift of the advancing blade, causes a torque ripple into the engine. Because of the finite bandwidth of the engine controller, the engine speed changes, resulting in changes in the shaft rate over time.

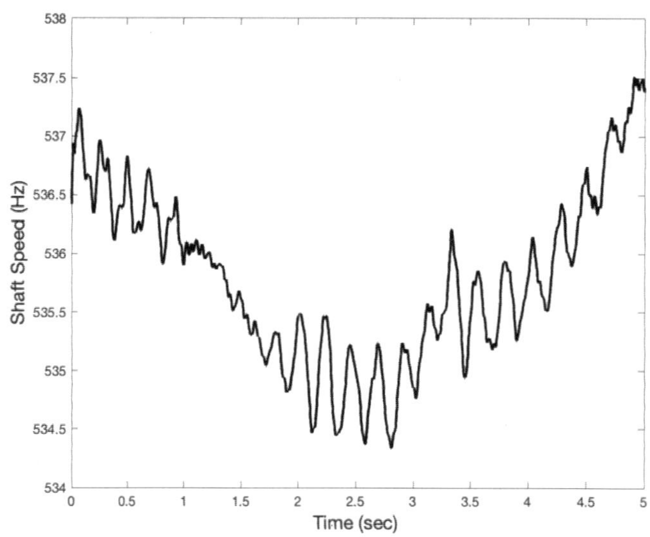

Figure 1 Shaft rate derived from the tachometer

These changes in shaft rate over time invalidates an assumption of stationarity when using the Fourier transform. This causes a smearing of the spectral content, as the energy associated with a shaft or gear is spread across many frequencies. As can be seen in FIG.1, the shaft frequency varies from 534.2 Hz to 537.5 Hz. If one were interested in the gear mesh frequency of this shaft (which has, as an example, 35 teeth), then for this range of shaft rates, the gear mesh frequency would range from 18,697 Hz to 18,813 Hz. The sample rate of the condition monitoring system may be, for example, 93,750 samples per second. Using the FFT, the length (i.e., the number of data points) of the analysis is typically some radix-2 value, for example, 16,384. Given this window length, then each FFT bin would have 5.722 Hz of frequency content. In this scenario, the frequency of the gear mesh is spread across 115 Hz or 20 bins. Hence, the representation of a feature would not be accurate. This is one of the reasons why gear and shaft analysis use the TSA, which removes the effect of shaft speed change by resampling the data (Ref 1, 5).

The TSA Algorithms: Analysis of the TSA was well described in Ref 9., and perhaps first implemented for gearbox analysis in Ref 3 (R.M Stewart went on the develop perhaps the first HUMS). As per Ref 9., the function to be averaged, $x(t)$, is digitized at a sample rate of nT, giving a waveform $x(nT)$, with an averaged period of mT, such that:

$$y(nT) = {}^{1}/_{N} \sum_{r=0}^{N-1} x(nT - rmT) \tag{1}$$

This is a linear filtering operation where it can be shown that the *discrete* transfer function is:

$$H(z) = {}^{Y(z)}/_{X(z)} = {}^{1}/_{N}\, {}^{1 - z^{-mN}}/_{1 - z^{-m}} \tag{2}$$

This is a *comb* filter, where the filter response is based on the number of averages, N, see **Error! Reference source not found.**. Note that the x-axis is in the order domain, where order 1 is the average, resampled shaft rate, and higher-order are the corresponding harmonic. Hence, for a 29-tooth gear, frequency content associated with the gear mesh would be present at order 29. This frequency not synchronous with the shaft would be filtered by, effectively, a FIR filter of length N (the number of averages), with coefficients $b_i = 1/N$. The attenuation would be based on the percent difference of the non-synchronous source. For example, say the shaft under analysis had a rate of 40Hz and was sampled for 16 revolutions. A nonsynchronous 44 Hz source would then have an order of 1.1, which would be attenuated by the TSA to 0.184 times the source level.

This clearly suggests that if the instantaneous frequency of, say a gear mesh, is somewhat different than the measured tachometer signal (due to jitter, or some gear dynamic phenomenology), the gear mesh signal can become attenuated by the TSA.

Practically speaking, the TSA is a simple algorithm. The TSA resamples the vibration associated with a shaft or gear, in the spatial domain, such that vibration associated with each shaft order, in the Fourier domain, represents one frequency bin. For example, in a system in which the shaft rate is such that for a given vibration sample rate, the acquisition system, on average collects 800 samples per revolution, the TSA would resample the 800 samples to 1024 data points (1024 is the next highest radix-2 value from 800).

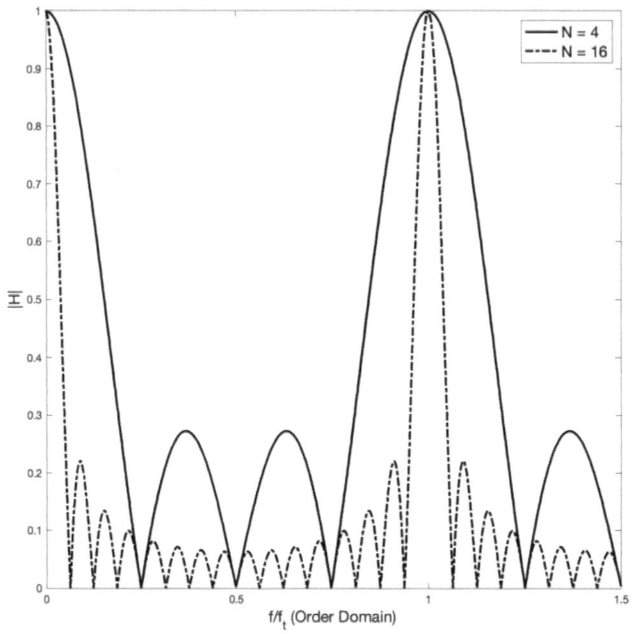

Figure 2 Comb filter response, order domain, for N = 4, and 16

Gear System Dynamics: Gears often use an involute profile to transfer the power, where the torque is transmitted through a conjugate action between mating tooth pairs. Mesh stiffness, which can be defined as the anti-deformation capacity of the gears under contact, varies significantly during a mesh cycle. The primary parameter which affects the change in mesh stiffness is the variation of the number of tooth pairs in contact. In addition to this, change in point of contact during a mesh cycle is also essential, since it is accompanied by a continuous change in contact stiffness of the tooth pair and change in individual bending stiffness of gears. A most significant mesh stiffness variation is observed for spur gear pairs. It is an important phenomenon for bevel gear dynamics as well. When such systems are excited near the critical speeds, the dynamic response can even increase to the levels such that loss of contact or even double-sided contact may occur, which is an indication of strongly increased dynamic forces (Ref 6). Note that this parametric excitation may become dominant not only at the fundamental mesh frequency but also at the harmonics of it due to the multi-harmonic content of the mesh stiffness.

This behavior can be modeled as a damped mechanic oscillator. A lumped-parameter, single degree of freedom (SDOF) gear dynamics model is provided in **Error! Reference source not found.** for a gear pair. Here *k(t)* is the time-varying mesh stiffness, *c* is the mesh damping constant, and *e(t)* is the external displacement action, which is known as gear transmission error representing the gear profile errors and/or intended modifications on gear tooth profiles.

Therefore, geared shaft systems have distinct dynamic characteristics. Such systems, as noted, have time-varying stiffness features, leading to a parametric excitation of the dynamic system.

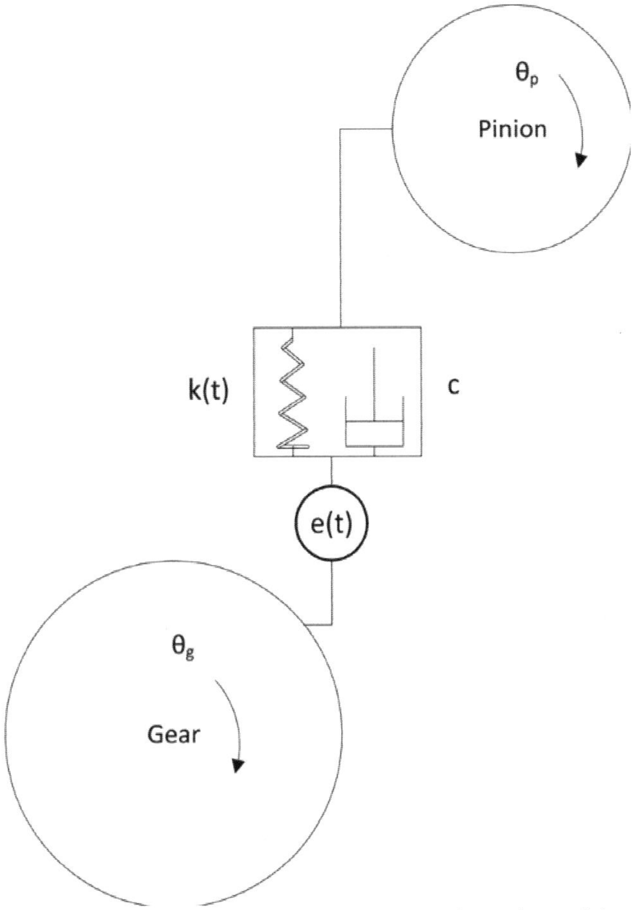

Figure 3 Lumped parameter 1 DoF gear dynamic model

Hence, the torsional dynamics of the geared shaft systems are not only affected by the shaft characteristics (i.e., geometry, material, etc.) but also from the gear meshes and their distinct dynamic features. Due to such complex dynamic behavior, it is not always possible to be away from every critical speed of the overall dynamic system with a wide margin. This also means system dynamics can be crucial in vibration-based drive train health monitoring. Coupled dynamic modes can also be existent where a dynamic phase difference is possible between any two rotating components of the system. This means pseudo-tachometer readings, which are derived based on gear ratios between the shaft with a tachometer and geared shaft of interest, may not always provide precise results for TSA calculations.

Such dynamic effects are suspected for the drive train under consideration in **Error! Reference source not found.**, where the TSA spectral energy of a gear mesh is considerably less than the measured spectral energy.

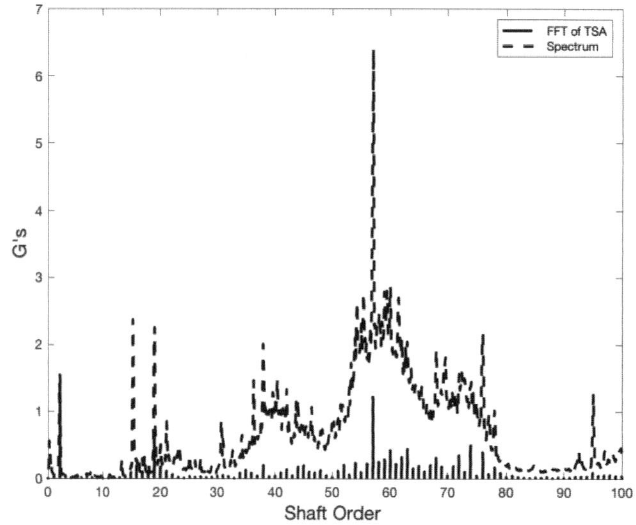

Figure 4 Raw vs. TSA spectrum

When a local tachometer signal was calculated from the vibration data, it is seen that the "local" tachometer signal shows a great deal of variation in rotation rate due to the dynamics of the system (**Error! Reference source not found.**). In general, the phenomena have not been observed because of limitations in the ability to determine the actual rotation rate of components "downstream" from the shaft being measured with a tachometer.

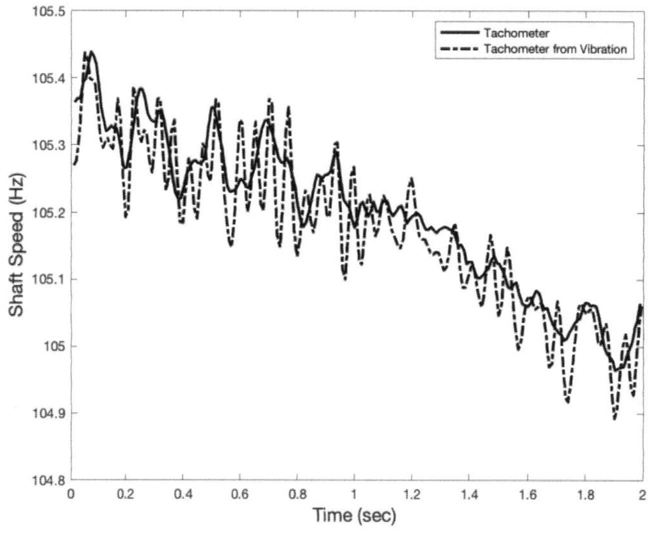

Figure 5 Comparison of tachometer derived shaft rate vs. vibration derived shaft rate

Deriving the Tachometer Signal form the Vibration:

Vibration signals from rotating equipment are sinusoidal and, by definition, synchronous with signals associated with the shaft rotation. However, the nature of vibration requires significant signal processing of vibration signals to extract useful information. Measured vibration is the superposition (i.e., addition) of many signals in the time domain. For example, consider a simple gearbox with an input shaft, an output shaft, and a gear pair. The input shaft turns at 30 Hz and has a 32-tooth gear, and the output shaft has an 82-tooth gear with a rotational speed of 11.707 Hz. The gear mesh frequency is 960 Hz (30 * 32). The gear mesh frequency will have sidebands as a result of any shaft imbalance being modulated onto the gear mesh. This can be proved using the trigonometric identity:

$$\cos(a) \times cos(b) = {}^{1}/_{2}\,[\cos(a+b) + cos(a-b)] \tag{3}$$

where, in this example, *cos(a)* is 960 Hz, and *cos(b)* is 30 Hz and/or 11.707 Hz shaft. Additionally, if the shaft is bent or bowed, there will be a 2x shaft vibration component. Other manufacturing defects, such as the gear not being mounted perpendicular to the shaft, or not centering the shaft on the gear (e.g., eccentricity), will result in additional frequency tones. To recover only those signals associated with the desired component, one can use an ideal bandpass filter and create an analytic signal in one functional procedure. This is followed by using a jitter reduction model to remove noise (jitter) from the reconstructed tachometer signal not associated with changes in the machine rate.

Recovering rotational information from vibration data involves using an estimate of the rotation rate of a component under analysis based on the tachometer measurement of a shaft and the shaft ratio to that component. From this information, a known gear mesh frequency for the component may be estimated. A range of frequency encompassing the estimated gear mesh frequency is found based on the variance in the measured shaft rate. By filtering around this range, the actual gear mesh frequency may be extracted from the overall vibration data even though its magnitude may be significantly smaller than the average overall vibration spectrum. Once the gear mesh frequency signal is determined, the actual shaft rate of the component of interest can be found. This may be accomplished, for example, by the following pseudo-code:

- Define the Sample Rate = sr. The number of data points, n, of vibration data equals *sr* x acquisition length in seconds, then:

- Calculate the next larger radix-2 length for the FFT: nRadix = $2^{\mathrm{ceil}(\log2(n))}$.

From the tachometer measurement and the gearbox configuration (i.e., the shaft ratio from the shaft measured by the tachometer to the shaft under analysis), calculate meta statistics such as approximate rotation rate of the shaft under analysis (i.e., the first moment), the variation (i.e., the second moment) in the approximate rotation rate, and the estimated known gear mesh frequency (also based on the number of teeth of gear on the shaft under analysis).

- From the estimated known gear mesh frequency and the variance in the estimated shaft rotation rate, calculate the low bandwidth index and the high bandwidth index (*bwlow*, *bwhigh*), which will encompass the gear mesh frequency of interest.

- Take the zero-padded FFT of the vibration data.

- Zero the FFT from zero to *bwlow*, and from *bwhigh* to nRadix.

- Take the inverse FFT to generate the analytic signal.

- Calculate the unwrapped argument of the generated analytical signal from to 1 to *n* time series. (The argument is the arctangent of the imaginary part of the analytic signal, to the real part of the analytic signal. Note that the value can only go from 0 to π and -π to 0. One is interested in the cumulative rotation of the analytic signal in time, hence as the signal exceeds -π to some small positive number, 2π is added to it. That is if at index *i* the value is -0.03π, and the next value calculated at index *i + 1* is 0.03π, the saved (unwrapped) value is 2.03π.)

- Normalize the time series of radians by the number of teeth of the gear (assuming 1st order harmonics).
- Interpolate the number of indexes for every 2π radians. The value 2π radians is one zero cross. Hence, the interpolation gives the exact index of the zero-cross of the shaft.
- Normalize to "tachometer" zero-crossing index by the Sample Rate (sr), which provides the zero cross time and from which the rotation rate for the component under analysis is calculated.

Idealized Bandpass Filter: A bandpass filter is the convolution of a low pass filter with a high pass filter. These filters are implemented as Finite Impulse Response (FIR) filters to improve their stability. However, even rather large filters have an inadequate response. Consider a case of a wind turbine gearbox, with an approximately 29 Hz shaft and a 32-tooth gear, which has a frequency of 928 Hz. The bandwidth of the filter is set from 910 Hz to 950 Hz to exclude the frequencies associated with the 30 Hz of the high-speed shaft (i.e., 29 * 32 - 29 = 899 Hz and 29 * 32 + 29 = 957 Hz). Using a large, 120 tap FIR filter, the bandwidth (at 50% power, 3dB) of this filter is 1670 Hz. The filter does not reject the spectral content at 500 Hz, or 1121 Hz, because the filter is not ideal and suffers from spectral leakage. These additional frequencies will be superimposed on the 928 Hz signal and will distort the desired analytic signal, reducing the quality of the resulting tachometer signal.
A process may be used in which developing the analytic signal using an ideal bandpass filter is completed in a single functional process. The analytic signal is defined for the real-valued signal $x(t)$, as determined:

$$X(f) = F\{x(t)\} \tag{4}$$

where F is the Fast Fourier Transform, and where:

$$X_a(f) = X(f),\ \ f = 0 \tag{5}$$

$$X_a(f) = 2X(f),\ \ f > 0 \tag{6}$$

$$X_a(f) = 0,\ \ f < 0 \tag{7}$$

$$x_a(t) = F^{-1}\big(X_a(f)\big) \tag{8}$$

where $X(f)$ is the Fourier transform of $x(t)$, and f is measured signal frequency.

Results: By reconstructing the tachometer signal from the local gear under analysis (Ref 7, 8), resulting zero-cross data improves the TSA, allowing for the full energy measured in the shaft under analysis to be used for gear analysis.

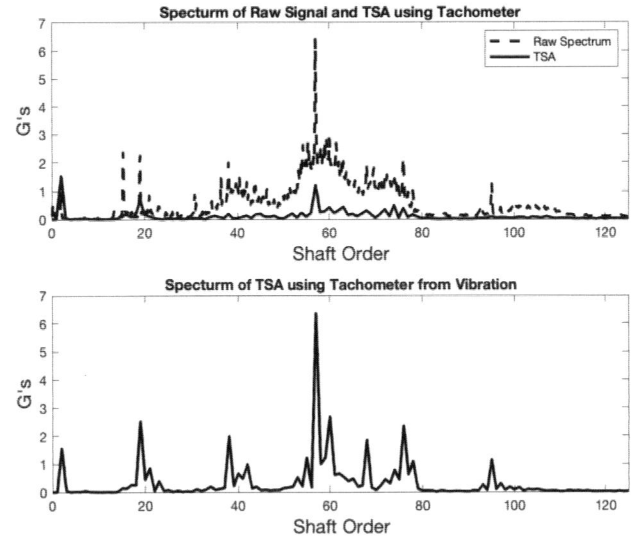

Figure 6 Improved TSA spectrum from improved analysis

Improving the signal to noise ratio using this technique will enhance fault detection. This can be seen in **Error! Reference source not found.**, comparing the raw spectrum and TSA spectrum (subplot 1), to the reconstructed tachometer TSA in Figure 1, subplot 2, were the SNR gain can be seen to be almost 8 dB.

This phenomenon has been observed for many gears in the main transmission of helicopters. For example, **Error! Reference source not found.**. shows the tachometer and reconstructed tachometer for the ring gear of a light single helicopter.

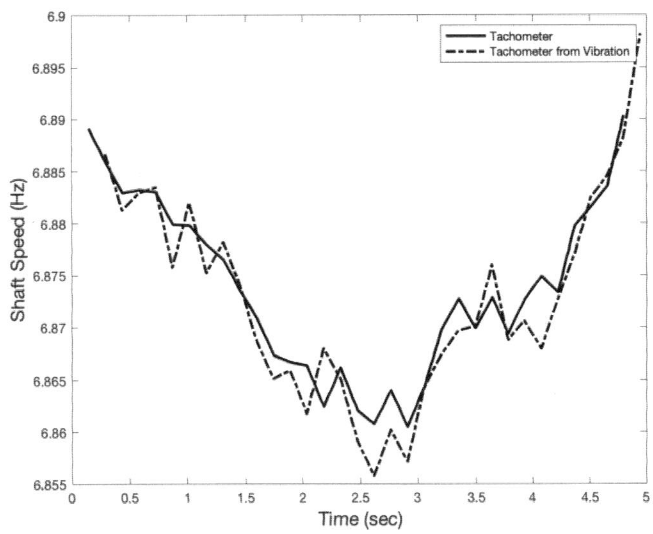

Figure 7 Tachometer vs. reconstructed tachometer signal, epicyclic

Note that because of the relatively high gear ratio of the main gearbox and resulting low speed of the main rotor, there are only 34 revolutions over the five-second acquisition. A comparison of spectrum using the TSA, vs. the "local" tachometer derived from the vibration signal, **Error! Reference source not found.** shows a 4-dB improvement in the signal then on the figure itself.

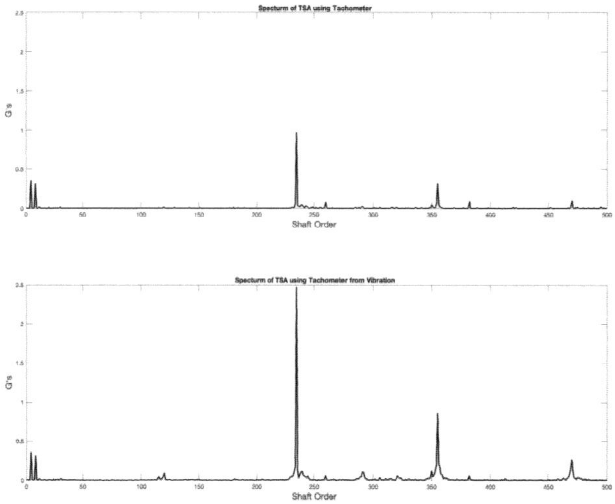

Figure 8 Comparison of the spectrum of the epicyclic gearbox using tach, and tach from vibration

Calculation of the Transfer Function due to Gear Dynamics: The transfer function, which represents the gear system dynamics, can be calculated by using the input signal (e.g., the measured tachometer signal) and the "local" tachometer signal. Consider that some transfer function H exists, then the local tachometer response is the convolution in time of H with the input tachometer signal:

$$zlocal(t) = h(t) * z(t) \tag{9}$$

Taking in Fourier transform

$$ZLOCAL(f) = H(t) \times Z(t) \tag{10}$$

Taking the log allows the nonlinear system to be expressed as a linear system. This technique is commonly known as homomorphic deconvolution.

$$log\{ZLOCAL(f)\} - log\{Z(t)\} = log\{H(t)\} \tag{11}$$

The transform function, for example, 1 (taken from the input and output of **Error! Reference source not found.**), is shown in **Error! Reference source not found.**. As seen from the phase change, the gearbox goes through resonance at approximately 17 Hz, as indicated by the phase change.

Conclusion: It has been observed in the TSA of some main gearbox analysis that the TSA spectrum had less energy than the raw spectrum at the gear mesh frequency. It was the hypothesis that gear system dynamics added phase error to the TSA for the "local" analysis, which degraded the TSA and resulting in an attenuation (due to filtering) of the gear mesh. This may adversely affect the resulting gear analysis. Using a novel, idealized bandpass filter, a local tachometer signal was derived for the shaft/gear under analysis, which significantly improved the

TSA, improving the signal to noise ratio. In real-world examples from a light single helicopter, a 4 to 8 dB improvement was observed in the TSA spectrum.

Improving the signal to noise of the TSA will improve fault detection for HUMS applications. Further, the ability to do analysis without a tachometer can help reduce the cost of HUMS hardware. While a tachometer 1/Rev striker is needed for main and tail rotor balance, in general, no phase information is required for drivetrain analysis. Besides the improved analysis, the allows in the main gearbox, removing a tachometer can reduce the HUMS kit cost, installation cost, and maintenance cost on HUMS itself.

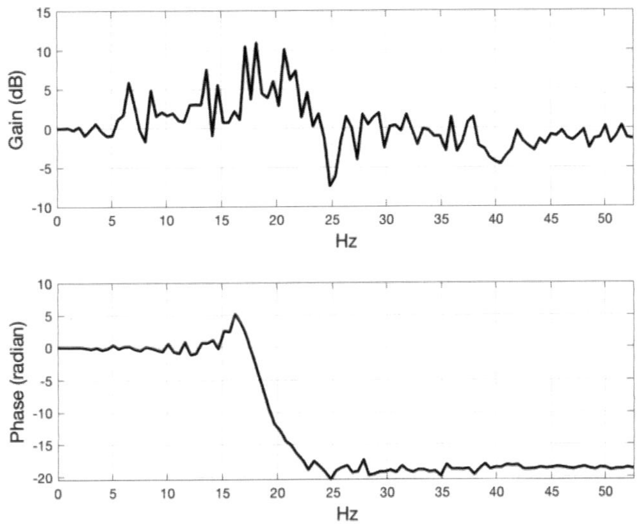

Figure 9 The measured transfer function due to gear dynamics

References:

[1] Bechhoefer, E., Kingsley, M, "A Review of Time Synchronous Average Algorithms", PHM Society Annual Forum, 2009.

[2] McInerny, S.A., Sun, Q., Hardman, B., "Public Domain Algorithms for Bearing and Gear Fault Detection," Proceedings of the 11th International Conference on Structure Borne Sound and Vibration, 2004s.

[3] Stewart, R.M., "Some Useful Data Analysis Techniques for Gearbox Diagnostics," Institute of Sound and Vibration Research, Proceeding of Meeting on Application of Time Series Analysis, Sept. 1977

[4] Bechhoefer, E., He, D., "Improving Gear Fault Detection by Reducing Tachometer Jitter," AHS Forum 72, May 1, 2014.

[5] Bechhoefer, E., Van Hecke, B., He, D., Processing for Improved Spectral Analysis, PHM Society Annual Forum, 2013.

[6] Ozturk, Y, "A Parametric Study on Planetary Gear Dynamics" Thesis, The Graduate School of Natural and Applied Sciences of Middle East Technical University, 2018.

[7] Bonnardot, F., El Badaoui, M., Randal, R. B., Daniere, J., Guillet, J., "Use of the acceleration signal of a gearbox in order to perform angular resampling (with limited speed fluctuations)," Mechanical Systems and Signal Processing 19 (2005) 766-785.

[8] Combet, F., Gelman, L., "An automated methodology for performing time synchronous averaging of a gearbox signal without speed sensor," Mechanical Systems and Signal Processing 21 (2007) 2590-2606.
[9] Braun, S., "The Extraction of Periodic Waveforms by Time Domain Averaging", ACUTSTIC, Vol. 32, 1975

Bibliography:

[1] Eric Bechhoefer is the president of GPMS, Inc., a company focused on the development of low-cost condition monitoring systems. Dr. Bechhoefer is the author of over 120+ juried papers on condition monitoring and prognostics health management and holds 23 patents and 5 patent applications in the field of CBM.

[2] Yalin Ozturk, is the Chief Engineer, Transmission Structural Analysis, at Turkish Aerospace. He is an experienced structural analysis engineer with a demonstrated history of working in the aviation and aerospace industry. Strong engineering professional with a PhD focused in Mechanical Engineering from Orta Doğu Teknik Üniversitesi / Middle East Technical University.

Improved Spectral Estimation of Signals using Quadratic Interpolation

Eric Bechhoefer[1], Brent Butterworth[2]

[1] *GPMS Inc, Cornwall, VT, 05753, USA*
[2]*Garmin International, Midlothian, TX, 76065, USA*

Abstract: Common in many fields of science and industry is the estimation of the magnitude and frequency of signals generated by periodic, sinusoidal signals. This process is ubiquitous and the foundation of many standards and procedures, and often involves use of the Fourier transform. In application, this is done with a Fast Fourier Transform (FFT). The FFT is defined only for frequencies that are exactly integers of the sample rate (*fs*). For other frequencies, this causes spectral leakage and error in the estimation of the FFT amplitude for frequency. On average, the magnitude error is 4%. This paper discuses an improved technique for the estimation of magnitude and frequency of a sinusoidal signal using the FFT, which reduces the magnitude error to a fraction of a percent. The improved estimation uses a transformation of the data to make the spectral peak parabolic, from which a quadratic solver is used to estimate the true magnitude peak value and frequency. This method can be proved to be optimal by reducing the FFT error in sum of square errors sense.

Keywords: spectral estimation, Fourier Transform, DFT, Optimization

Introduction: In many fields of science and industry, there is a need to estimate the magnitude and frequency of signals generated by periodic, sinusoidal signals. This process is ubiquitous and the foundation of many standards and procedures, and usually involves the Fourier transform. For example, in utility power generation and distribution systems, i.e., the electrical grid, it is common for line frequency to be run at 60 Hz. However, this frequency is actually 60 Hz +/- some small percentage error. In order to make sure the frequency stays within acceptable margins, power providers must determine both the frequency and magnitude of the voltage and the current of electricity being generated at power plants. An accurate way to determine these parameters involves performing a Fourier transform on the measured data.

Similarly, the standards for both the International Electrotechnical Commission (IEC) and the Environmental Conditions and Test Procedures for Airborne Equipment (DO-160) require taking measurements of electromagnetic radiation related to the control of emission and transmission of radiant energy from equipment. These measurements are typically made using spectrum analyzers that incorporate Fourier transform.

Likewise, in vibration-based condition monitoring, signals relating to rotating components of machines are monitored so that mechanical diagnostics of the equipment may be performed. Periodic signals associated with any nonconformity of the rotating components, such as shaft imbalance, gear faults, and bearing faults, are predominantly quantified using spectral analysis. This spectral analysis is usually done by measuring vibration with a calibrated accelerometer, the analog signal from the accelerometer is digitized and an analysis is performed that uses Fourier transforms in order to measure the magnitude of vibration at a given frequency. The frequency is associated with the properties of the rotating component. For example, a wind turbine may have a high-speed shaft driving a generator at 1800 RPM (or 30 Hz). If the shaft has a 24-tooth pinion, the frequency associated with this gear is 24 x 30 Hz, or 720 Hz. The energy associated with this frequency can be quantified using the Fourier transform.

In another example, for the evaluation of the performance and load of an induction motor, it is common to calculate slip. Slip is the percent difference between the synchronous speed (i.e., line frequency) and the speed of the motor. Excessive slip indicates the load on the motor is beyond the design limit. It is usually assumed that synchronous speed is 50 or 60 Hz, but it is well known that at any given time the line frequency has error. Taking the Fourier transform of the line current (using a current transformer), will give a more exact measure of the synchronous speed. Measuring the shaft order 1 vibration with an accelerometer and then taking the Fourier transform can make a more exact determination of the motor speed. Thus, estimating slip can be improved using the Fourier transform.

In addition, many systems or tools that measure periodic signals need to be calibrated to ensure accuracy. For example, it is common in the field of condition monitoring to test accelerometers with a handheld shaker (such as Model #394C06 available from PCB). The shaker has an acceleration output of 1.00 G's (+/- 3%) and an operating frequency of 159.2 Hz (+/- 1%). In order to ensure compliance with any specification, it is important to accurately measure both the magnitude and frequency of the shaker. This analysis may be validated using the Fourier transform of the measured signal.

Taking exact Fourier transforms can be impossible and is often impractical, so for most applications a discrete Fourier transform (DFT) is implemented. Because of the high computational load often required even for DFT, a Fast Fourier Transform (FFT), which is an algorithm that computes the DFT of a sequence, or its inverse, is typically used. As noted, the FFT for some frequencies can have error: in many applications, there is a need for an improved FFT technique that can attain greater levels of accuracy for magnitude and frequency of measured data. Of course, one would like to do this without significantly increasing computational load.

Fast Fourier Transform: In the context of signal data analysis, the Fourier transform is used to decompose a time domain signal, which is the superposition of frequencies, into the frequencies that make up the signal. In implementation, the Fourier transform is calculated using the Fast Fourier Transform (FFT) algorithm. The FFT makes an assumption that the measured signal is a finite sequence of equally spaced measurements. The bandwidth, or Nyquist frequency of the FFT, is half of the sampling rate (SR) of the discrete signal. The frequency resolution of the FFT is then the SR divided by the length, N, of the FFT, or SR/N.

A key property of sinusoids is that are orthogonal at different frequencies: $A_1 sin(\omega_1 + \phi_1)$ is orthogonal to $A_2 sin(\omega_2 + \phi_2)$. For length n sampled sinusoids signal, such as used in the FFT, this property of orthogonality only holds for harmonics of the sample rate divided by n, or specifically, for frequencies;

$$f_k = k\, {}^{f_s}\!/_{N}, \quad k = 0,1,2,\dots,N-1. \qquad (1)$$

As such, the FFT is defined only for frequencies that are exactly integers of the sample rate (fs), e.g. ωk = 2πkfs/N. For frequencies that are not exactly integer of the sample rate, it can be shown that the FFT acts as a digital filter, where the frequency response for some frequency response for some unknown frequency, ωk, is:

$$|X(\omega_x)| = \left| \frac{sin\{(\omega_x - \omega_k)\, NT/2\}}{sin\{(\omega_x - \omega_k)\, T/2\}} \right| \tag{2}$$

which causes spectral leakage and error in the estimation of the FFT amplitude for frequency ωk. Note that spectral leakage is not minimized by increasing N.

While there are a number of different algorithms for the FFT, many are based on the Cooley-Tukey method, which is a sort of "divide and conquer" algorithm that recursively breaks the length of each data point into a composite of N = N1 x N2, where N is the composite size and N1 and N2 are smaller factors. The best known of these algorithms are based N2 = 2, which results in a radix-2 algorithm. For these algorithms, the total length (N) of the FFT is necessarily a 2n power, for example 4096 (2^12) or 8192 (2^13), see Smith (2007).

Spectral Density Estimation: One issue that sometimes needs to be addressed when using FFT for data analysis is the Gibbs phenomenon, which is observed in the Fourier series of a piecewise periodic function that has a jump discontinuity in a signal and occurs with finite data lengths. A method used to control the Gibbs phenomena and therefore obtain a better estimate the power of a signal is Welch's method. Welch's method splits the signal into, typically, radix-2 lengths (the window length) and overlapped segments (the window length divided by two). The overlapped segments are then windowed, and a window function mitigates the Gibbs phenomena. A typical window function, applied to the segment, is the Hann window. The Hann window is smooth, and bell shaped, and thereby mitigates the edge/jump discontinuity of a segment on which the FFT operates.

Welch's method returns the power of a signal in each frequency bin. The magnitude of the signal is then the square root of the power of the signal in a given frequency bin. The frequency resolution of each bin is a function of the sample rate and length to the window of Welch's method. For a more complete treatment, Smith (2007).

A Real-World Application: As an example, consider the calibration of a handheld shaker. The acquisition system samples an accelerometer to measure voltage by an analog-to-digital converter (ADC). The ADC sample rate is based on times fixed by a clock or a crystal oscillator. Our oscillator device has a frequency of 12 MHz +/- 12 parts per million (ppm). The ADC sample rate is then the clock frequency divided by some radix-2 power. For example, a power of 11 would give an ADC sample rate of 12,000,000/2048, or 5859 samples per second. Given the SR of 5859 samples per second and a Welch's window of 4096, the resolution of each FFT frequency bin is 1.4304 Hz.

If the frequency of the handheld shaker were exactly 159.2 Hz, then the exact frequency bin on the FFT would be the 111.296 bin index (i.e., 159.2/1.4304). Unfortunately, the resolution of the FFT, as previously shown, is finite, therefore some of the energy in the 159.2 Hz acceleration is found in bin 111 (corresponding to a frequency of 158.776 Hz), and some is found in bin 112 (corresponding to a frequency of 160.21 Hz), as can be seen in figure. 1. This is the spectral leakage. Note also in figure. 1 that at the maximum magnitude is 0.944 G's, which translates to a 5% error for the measured 1 G input signal for the handheld shaker. In this example, an error of that amount would mean the handheld shaker would fail the calibration.

In general, a strategy for improving the calibration process may be to try to increase the resolution by either reducing the sample rate or increasing the window length of the FFT. If the window in the above example were increased to 32768, then the bin index would become 890 (159.2/(5859/32768)). This corresponds to a frequency of 159.13, which is closer to true frequency, but the magnitude is then only 0.915 G's because more energy is found in bin index 891. At noted earlier, increasing N does not reduce spectral leakage.

Spectral Interpolation: We will use the shaker calibration as an example to demonstrate a better estimate of the FFT spectrum magnitude and frequency of a measured signal of 1g at 159.2 Hz. The sample rate is 5859 with a window of 4096, in this example. The maximum energy is likely found at a frequency between 158.8 Hz and 160.2 Hz.

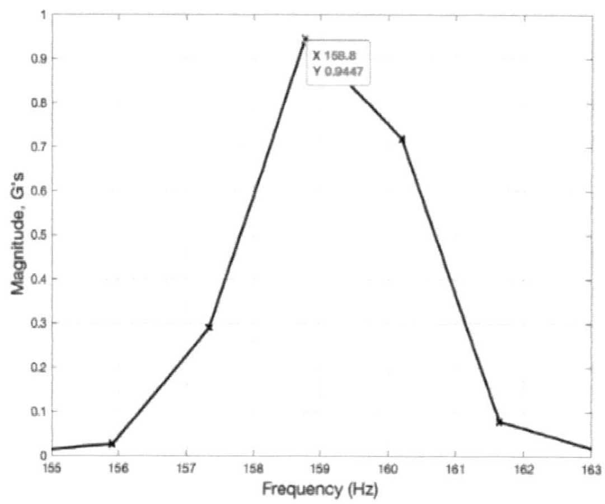

Figure 1 FFT of a 159.2 Hz sinusoid at 1 g

Further, there is a significant amount of energy at 157.3 Hz. Interpolation can allow for an estimate of the maximum energy value and the frequency at which it occurs by assuming a continuous function or distribution. A polynomial quadratic, which is a smooth function, can be used to interpolate the maximum energy through, for example, a triplet of points.
A polynomial quadratic has a general formula that can be expressed as:

$$y(x) = a(x - p)^2 + b \tag{3}$$

where *a* is the curvature and depends in this context on the window used in the FFT; *p* is the center point and gives the interpolated peak location; and *b* is the amplitude that here equals the peak amplitude of the FFT spectrum. The three samples nearest the peak can be defined as:

$$y(-1) = \alpha,\ y(0) = \beta,\ \text{and}\ y(1) = \lambda,$$

where the bins about those three peaks are defined at [-1 0 1]. Substituting the bin values for x and simplifying results in the following expressions for α, β, and λ:

$$\alpha = ap^2 + 2ap + a + b \tag{4}$$

$$\beta = ap^2 + b \tag{5}$$

$$\lambda = ap^2 - 2ap + a + b \tag{6}$$

By combining equations, substituting, and rearranging terms, the following relations can be derived:

$$\alpha - \lambda = 4ap \tag{7}$$

$$p = {(\alpha - \lambda)}/{4a} \tag{8}$$

$$\alpha = ap^2 + \left({(\alpha - \lambda)}/{2}\right) + a + (\beta - ap^2) \tag{9}$$

$$a = {1}/{2}\,(\alpha - 2\beta + \lambda) \tag{10}$$

Then, the interpolated peak location, *p*, in bins, can be expressed as:

$$p = {1}/{2}\,\frac{\alpha-\lambda}{\alpha-2\beta+\lambda} \tag{11}$$

If *k* is the index of the maximum spectral value, then the interpolated frequency (*frq*) would be determined as follows:

$$frq = (k + p)\,{SR}/{window\ length} \tag{12}$$

The interpolated magnitude (*mag*) could be determined from the following:

$$mag = y(x) = \beta - {1}/{4}\,(\alpha - \lambda)p \tag{13}$$

Again, see Smith (2007).

In the example data from Figure 1, the three points at 157.346 Hz, 158.776 Hz, and 160.21 Hz, having energies of 0.2895 G, 0.9447 G, and 0.7184 G, respectively, are used to interpolate the maximum energy. As shown in Figure 2, these three points serve as the points through which the polynomial is fitted, and are used to find values for α, β, and λ for the data. With no further transformation of this data, the above interpolation technique results in a peak value of 0.9708 G with a frequency of 159.1246 Hz.

This estimation of the magnitude and frequency is based on the interpolation without any transform of the original spectral data. However, a Gaussian window magnitude is precisely a parabola on a dB scale, which implies that use of the above quadratic interpolation technique is exact in the dB scale. If this were the case, then logarithmic transformation that defines the relationship of three points in the spectrum (here, α, β, and λ) would minimize the error between the true magnitude and frequency and the interpolated magnitude and frequency.

A test for such a transform would be bounded by the minimum error between the points of the interpolation and those points tested by the transform. As noted, the error is zero where the FFT bin is exactly the frequency of the sinusoidal signal being processed. For the given example with a sample rate of 5859, a window of 4096, and a desired frequency of 159.2, the exact frequency of zero error (at bin 111) would be 158.7766 Hz (111 * 5859 / 4096). The bound from bin 111 to bin 112 would be a frequency of 160.2070 Hz (112 * 5859 / 4095). The maximum magnitude error would then occur at 111.5, or a frequency of 159.4918 Hz.

A log transform is one option: the log transformation tends to compress the peak values, so other potential transformations include the square root, the cube root, the fourth root, and the fifth root.

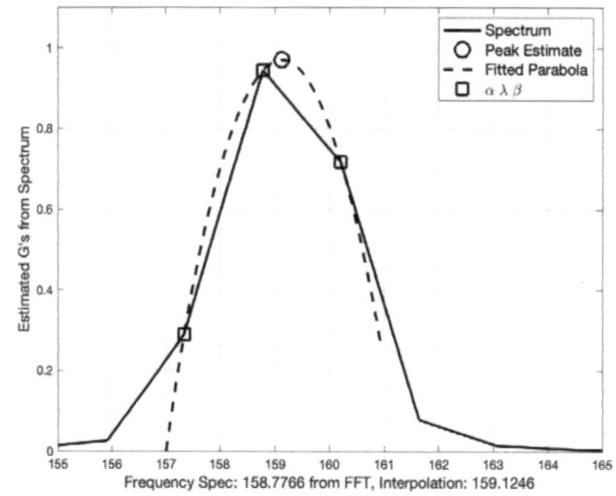

Figure 2 Polynomial Estimation of 0.971 Gs at a frequency of 159.12 Hz

Figure 3 is a graph of the frequency determined by interpolation and the associated percentage error that includes results for the raw FFT data and for estimates made via the parabolic interpolation based on linear, log, square root, cube root, fourth root, and fifth root transforms. The raw FFT results show greater error compared to any of the parabolic estimates. Note that in the zoomed plot, the minimum frequency error is between the 4th and 5th root, and that the log transform underestimates the frequency.

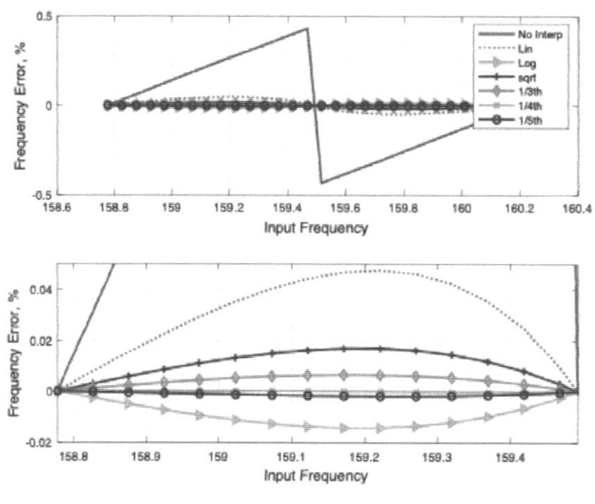

Figure 3 Frequency Error in percent for various transforms

Figure 4 is a graph of input frequency and the associated magnitude percentage error for the raw FFT, and for parabolic estimates based on the linear, log, square root, cube root, fourth root and fifth root transforms. A maximum error of 14% in the raw FFT is large in this context. Again, as seen in the zoomed plot, the minimum error is again between the 4th and 5th root. Note also that the magnitude errors are typically much larger than the frequency errors.

Optimization to Find the Best Transform: For figures 3 and figure 4 it is seen that there must be some transform of the data, between the 4th and 5th root, that minimizes the percent error in the spectral estimation. This root can be found as an optimization problem, where the goal is to minimize some objective function or score that represents the global error. One such objective function is the sum of square errors. That is, over the frequency in interest, for each trial transform, the performance metric will be the sum of square error over the bounded frequencies. An optimization technique, such as Newton-Raphson method (Press, 1998) or some other optimizer algorithm can be used to find the optimal root. In the given example, the optimal root is determined to be the 4.3093th root.

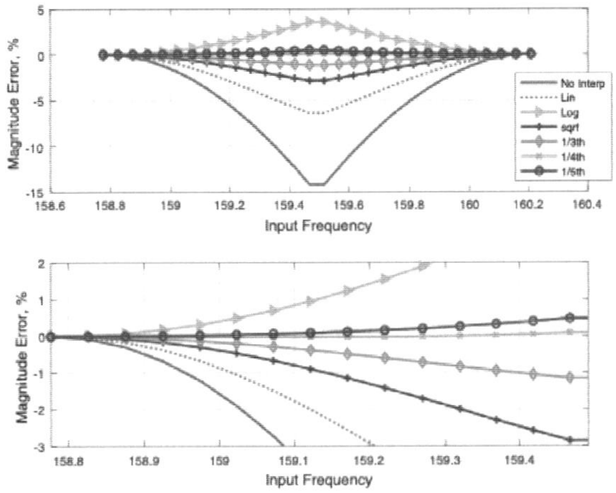

Figure 4 Magnitude Error in percent for various transforms

An Optimization Method: The optimal transformation is determined by finding the root that minimizes the percent error for an idealized function that produces a signal with an amplitude of one. Since this approach is based on minimizing percent error, it is generalized. For example, this test can be made using a sample rate of 5859 samples per second with four seconds of data, using an FFT window in Welch's method of 4096 with an overlap of 2048. The signal of interest is, for example, 159.2 Hz. From DFT theory, each FFT bin is the sample rate divided by the window, which here is 1.403 Hz. As shown, the exact bin for the frequency would be 159.2/1.403 or bin 111.296. However, since the bin is an integer, the frequency will fall between bin 111 and bin 112.

To solve and find the best transform, a sample of idealized frequencies is complied, from bin 111 (111*1.403 = 158.7766 Hz) to 112 (160.2070 Hz). Partitioning the space 50 times, for a given test frequency (*crtfrq)*, which creates an array that is 23436 in length. Since each bin to bin + 1 is the same size, the accuracy of an FFT can be tested for a known transform with a known signal source, s, which has an amplitude of one. Thus, the magnitude of any trial (i.e., the output of the FFT) should be one, and the frequency of the signal source is denoted as *crtfrq*. The following pseudo code determines *s* and *crtfrq*:

```
step = sr/win/50
idx = 1, 2, 3, ... 5859 x 4
for i = 1:50
      crtfrq = 158.7766 + step * i;
      s = sin(idx / sr * 2 * pi * frq);
```

Then the optimal root can be determined by minimizing the percent error in the amplitude, which can be accomplished as follows:

```
j = 30;
sr = 5859;
n = 5859*4;
win = 4096;
df = sr/win;
idx = 159.2/df;
start = df*floor(idx);
en = df*round(idx+1);
frq = linspace(start,en,j);
scale = 1;
for i = 1:j
      s=sin((1:n)/sr*2*pi*frq(i))*scale;
      [mg(i),fq(i)] =
    quadInterSpecPeakOps(s,win,root);
end

me1 = -(scale-mg)/scale * 100;
mse = mean(me1.^2);
```

Where the interpolation function is given as (in Octave or Matlab®):

```
function [mag,freq] = quadInterSpecPeakOps(v,win, root)
%[mag,freq] = quadInterSpecPeak(v,sr)
%v  :is the signal
%sr :is the sample rate
%quadratic interpolation of the spectral peaks
sr = 5859;

[spc,~] = pwelch(v,win,win/2,sr);
[~,idx] = max(spc);
abg = spc([idx-1 idx idx+1]);
abg = nthroot(abg,root);
p = 1/2*(abg(1)-abg(3))/(abg(1) -
 2*abg(2) + abg(3)); %interp peak bin
freq = (idx+p-1)*sr/win;
mag = abg(2) - 1/4*(abg(1)-abg(3))*p;
mag = mag.^root;
```

In Figure 5, the frequency error for the magnitude estimations based on the fourth root, the fifth root, and the determined optimal root of 4.3093. Note the error is less than 0.5e-3%.

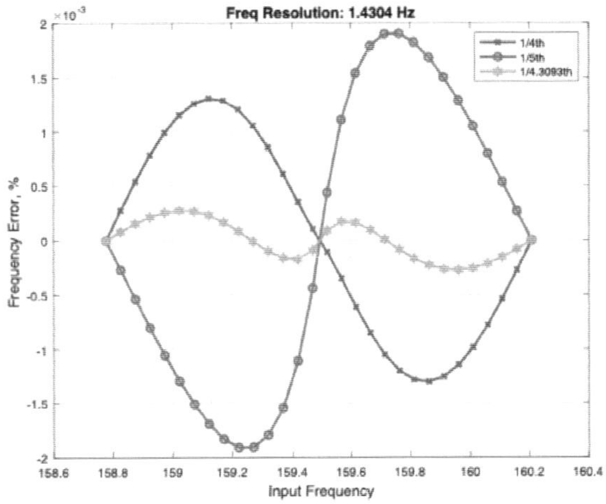

Figure 5 Frequency error based on optimization of the 4.3093th root

In Figure 6, the percent error for magnitude estimates based on the fourth root, the fifth root, and the determined optimal root of 4.3093. The optimal root of 4.3093 was robust in that is did not change with changes in window size, such as from 8192 to 16384 to 32768. The maximum error was less than 0.07%.

Table 1 compiles the: maximum percent magnitude error, maximum percent frequency error, and mean magnitude error for estimations in the above example based on the various transformations used as part of the interpolation.

Transform	Max. Mag. Error	Max. Freq. Error	Mean Mag Error
RAW FFT	14.16%	0.433%	-4.99%
Linear	6.37%	0.047%	-2.48%
Log (dB)	3.58%	0.014%	1.27%
4th Root	0.18%	0.001%	-0.09%
5th Root	0.48%	0.002%	0.15%
4.3093th Root	0.06%	0.0003%	-0.009%

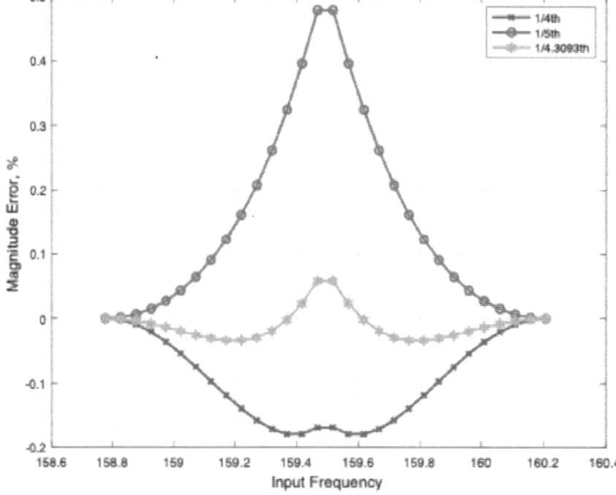

Figure 6 Magnitude error for the optional root of 4.3093

As can be seen from the values in Table 1, the magnitude estimation determined using the optimal root (in this example, the 4.3093th root) transform derived from the present invention is on average 190x lower error compared to the log transform with a frequency estimation that is 60x times lower percent error than the log transform. Using the optimal root for the transform thus greatly improves the mean and maximum error of the parabola interpolation. This in turn improves the value of spectral analysis based on the FFT.

Conclusion: It is surprising that given the ubiquitous usefulness of the FFT, that there is not more concern about the accuracy of the derived magnitude and frequency of the results. In general, the FFT provides a good estimate, but as noted, on average has an error 4%. Depending on the application, this may be unacceptable.

Because the improved estimation of magnitude and frequency does not require significant additional computational load, this methodology should be integrated into any number of applications, including condition monitoring where determining the energy associated with a bearing energy, or other asynchronous processes, is common. It must be noted that with synchronous averaging, each bin represents an exact frequency harmonic, such that this interpolation is not needed (Bechhoefer, 2009).

References:

[1] Smith, J., (2007). *Mathematics of the Discrete Fourier Transform (DF) with Audio Applications.* Book Surge, ISBN 978-0-97545607-4-8

[2] Press, W, Flannery, B, Teukolsky, S., Vetterling, W., (1988), *Numerical Recipes in C.* Cambridge University Press.

[3] Bechhoefer, E., Kingsley, M., "A Review of Time Synchronous Average Algorithms", Annual Conference, of the Prognostics and Health Management Society, 2009

Bibliography:

[1] Eric Bechhoefer is the president of GPMS, Inc., a company focused on the development of low-cost condition monitoring systems. Dr. Bechhoefer is the author of over 120+ juried papers

on condition monitoring and prognostics health management and holds 23 patents and 5 patent applications in the field of CBM.

[2] Brent Butterworth has been in Aviation Rotorcraft Maintenance and Engineering for 25+ years. He is a veteran of 10 years in the USMC and USAF. He held various leadership roles within Airbus Helicopters Technical Support and Engineering group. He is currently a Systems Engineer at Garmin in the G5000H architectures. He received his B.S. in Professional Aeronautics from Embry Riddle Aeronautical. He holds an FAA A&P license, FCC GROL with Radar Endorsement and Amateur Radio License. He is currently pursuing a B.S. in Data Management and Analytics.

*** Formatting Issue - Best Available Paper/Graphic ***

RIGHT-SIZING OF TURBO-GEAR BOX FOR CENTRIFUGAL COMPRESSOR WITH SYNCHRONOUS MOTOR AS DRIVER

Mantosh Bhattacharya, Nirav Doshi

Petrofac International, Sharjah, UAE

Mantosh.b@petrofac.com , Nirav.Doshi@petrofac.com

Abstract -

In oil & gas industry, for high speed centrifugal compressors speed increasing gear boxes are used to achieve high polytrophic head and flow. API standards and project specifications (as addendum to API) are the guidelines used to design compressor trains.

The proposed paper presents a study in order to stimulate a thought process for end users / purchasers to prudently use API 613, 5th edition standard, where gear box sizing can be hugely impacted by driver motor rating if due diligence is not applied .

The ramifications of oversized gear box may end up with longer / very expensive coupling larger footprint of the train, a bigger motor, with larger lubrication oil requirements and oil cooler size.

A detailed internal study including sensitivity analysis of subject gear box with designed centrifugal compressor, for varied operating process conditions reveals that incorporation of such a clause into the specification may not be very fruitful for defined operating parameters

The paper finally concludes that bulleted API clause should be prudently applied on specifications based on well production and depletion profile with a future plan of bundle replacement of the compressor.

Key words- Synchronous motor, Helical gear box rating: torsional ripples: Flexible coupling: Compressor configuration.

Introduction

In Oil & Gas, refinery and chemical industries, high speed centrifugal compressors are used very frequently with speed increasing double helical gear units. In design of these centrifugal compressor trains, API (American Petroleum Institute) standards and Project specifications (which are normally addendum to API) are followed.

Speed increasing double helical gear boxes are selected to conform to API 613 latest standard and end user specifications (addendum of API 613). As a good engineering practise, gear ratio is normally kept below 10. API 613 calculations for the vital features of a gear unit design, such as tooth pitting index and bending stress, are based on well proven AGMA methods with some additional factors which are related to meet extended service life of the gear box. The capacity of gears to transfer a given torque for a desired operating life is determined by computing the strength

*** Formatting Issue - Best Available Paper/Graphic ***

of the gear teeth in bending and the durability of the teeth (i.e., resistance to wearing / bearing/ scuffing loads). AGMA 6011-J14 provides details about the strength and durability calculations.

Fig. 1: Double helical gear set internals (bull gear & pinion)- (courtesy Renk)

API 613, 5th edition, bullet clause 2.2.1 states as, "The gear unit rated power shall be specified by the purchaser. For gear units located next to the driver, the maximum gear unit rated power shall be the maximum installed power of the driver. For electrical motor drivers, the gear unit rated power shall be the motor nameplate rating multiplied by the motor service factor". Bullet at the beginning of the paragraph indicates that either decision is required, or further information is to be provided by Purchaser. This information should be indicated on the data sheet, otherwise it should be stated in the purchase order.

Discussion and deliberation to enforce above clause in project / purchase specification for an electric motor driven centrifugal compressor unit is the primary objective of this paper. The paper demonstrates how centrifugal compressor design can have an influence on the above clause.

Let us consider a centrifugal compressor required by end user with desired process parameters and subsequent design details as below with an example –

Centrifugal compressor Inlet condition:	**Centrifugal compressor outlet condition:**	**Compressor design detail:**	
Pressure (bara): @18	Pressure (bara): @79	No of impellers	6
Temperature(°C): 50	Temperature(°C): 165	Rotor Speed (RPM)	13500
Mass flow (kg/s): 27	Polytropic head (KJ/Kg): 190	Rotor configuration	Beam type
Inlet volume(M3/hr):6400	Required BKW – 8.8 MW	Installed Motor KW	10.75MW

*** Formatting Issue - Best Available Paper/Graphic ***

Driver sizing is based on mechanical losses in seal system, lubrication system and mechanical efficiency of gear box and motor itself. Motors operating at higher ambient temperature are dealt with proper derating factor. So, motor installed horsepower is always higher than calculated brake horsepower at drive end coupling of gear box / compressor.

A speed increaser double helical gear box (as per API 613) with speed ratio of 9 is placed between driver and compressor to meet higher polytropic head requirement between driver and compressor with following basic geometry:

Bull gear teeth	217
Pinion teeth	24
Helix angle	25°
Pressure angle	20°

Two design options are being postulated here in this paper for gear box with different rating

Option I:

Double helical gear box with gear ratio of 9 with mechanical rating of 10.75 MW (Based on main motor name plate rating in order to comply the API 613 bullet point number 2.2.1 & project specification requirement).

Option II:

Double helical gear box with gear ratio of 9 with mechanical rating of 8.8 MW (Based on maximum compressor power required at worst case scenario considering threshold of choke flow).

Detail discussion on Option I and its possible limitation:

In order to just comply the API 613 and its modified project specification, in case Option I is selected then followings issues may be faced:

1. Bigger gear box:
    a. As gear box diameter increases, its manufacturing become more complex and need special care. Opting forged design for bull gear for higher reliability hence with bigger bull gear diameter, forged design will lead to higher material cost as well high manufacturing cost & high precision required.
    b. Bigger bull gear diameter will increase the gear box overall width which leads to increase in base frame width consequence a bigger foundation.
    c. Bigger gear box will have larger shaft and journal load leading to higher power losses and, may increase oil requirement (which lead to bigger lube oil system).

***** Formatting Issue - Best Available Paper/Graphic *****

Refer below comparison table between 8.8MW & 10.75MW gear boxes:

Features in comparison	**Gear Box (10.75MW)**	**Gear Box 8.8MW**
Shaft Diameter Bull gear (mm)	223	223
Shaft Diameter pinion (mm)	123	123
Pitch Diameter Bull gear (mm)	1670	1445
Pitch Diameter pinion (mm)	170	150
Shaft center distance (mm)	900	790
Face width (mm)	270	300
Material property - Bending stress (MPa)	210 (Pinion) / 205 (Bull Gear)	208 (Pinion) / 203 (Bull Gear)

2. Torsional stability of unit having various type of electric motor drives: Large inertia constitute a weakly damped mechanical resonator exhibiting a relatively low resonance frequency which get active during unit start up.

Below paragraphs provide short notes on torsional vibratory (oscillation) characteristics are provided for synchronous motors as turbo compressor drivers.

Synchronous Motor drive -

Unlike induction motors, synchronous motors are not self-starting, and they are normally equipped with squirrel-cage windings which provide starting torque and also provide damping during steady state running. The stator's magnetic field begins rotating at synchronous speed virtually instantaneously after power is applied. With the stator field rotating and the rotor at rest, alternating forward and reverse torques are applied to the rotor. This causes the rotor to swing back and forth These windings are utilized to accelerate the motor as an induction motor from zero speed to a speed slightly less than synchronous speed. When synchronous speed is approached, DC field voltage is applied, and the rotor is pulled into synchronism. It is known that synchronous motor rotors contain salient poles that are magnetic protrusions enclosed by field coils.

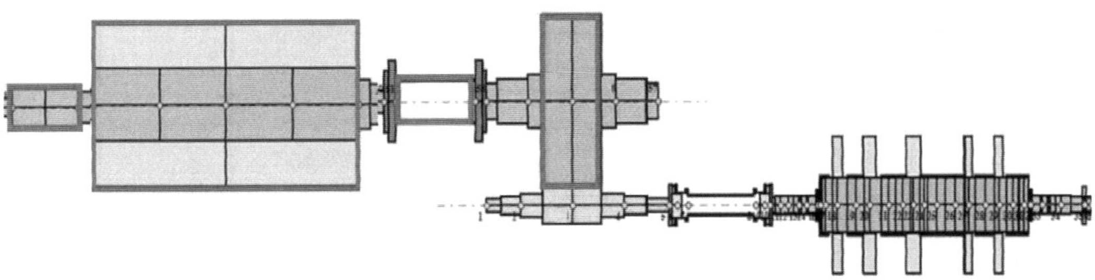

Fig 2 – Synchronous motor drive train torsional model., (courtesy Dyrobes)

***** Formatting Issue - Best Available Paper/Graphic *****

The frequency of the torque pulsations is the frequency at which the stator's rotating magnetic field passes a rotor pole. Since the stator's magnetic field rotates at synchronous speed, the excitation frequency is a function of the difference between synchronous speed and rotor speed, which is known as slip speed. Specifically, excitations occur at twice slip frequency where slip frequency is defined by the following equation:

$$F_{slip} = FL * (Ns - N) / Ns$$

where:

Fslip is Slip frequency (Hz)

FL -Line frequency (Hz)

Ns -Synchronous speed (rpm) Rotor speed (rpm).

N - actual Rotor speed at that point of time.

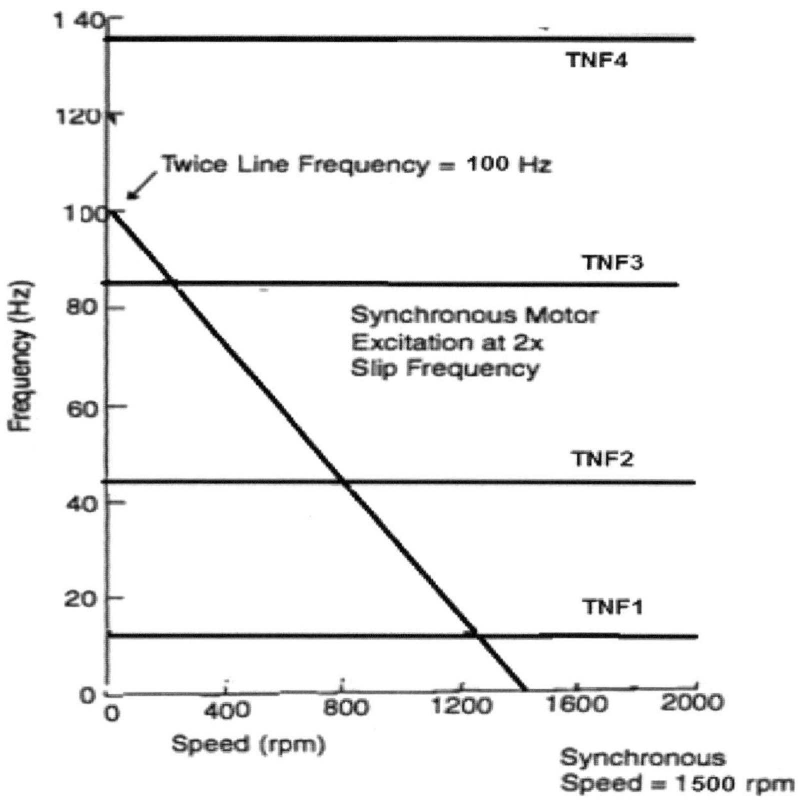

Fig 3– Synchronous motor start up diagram

All synchronous motors produce torque pulsations between start and synchronous speed. For 60 cycle power synch motor that start inductively have pulsing torque starting at 7200 cpm and decreasing inversely with speed to 0 cpm at synchronization. The shaft/coupling will normally

*** **Formatting Issue - Best Available Paper/Graphic** ***

have a torsional resonance in that frequency range and cause gearing noise through the critical. It will always occur over a fairly narrow speed range and will repeat from start to start. It will not be excited on coast down since the torque pulses only occur on start. For 50 hz power supply frequency the pulsation frequency will go from 6000 to 0 cpm on start These pulsations start at twice line frequency (120 Hz or 100 Hz) and decay to zero Hz when the motor synchronizes. Thus, all torsional resonances between 0 and 7,200 CPM /6000 CPM will be excited every time the motor is energized. These torque pulsations can be several times the rated full-load torque of the motor and are known to cause major machinery failures.

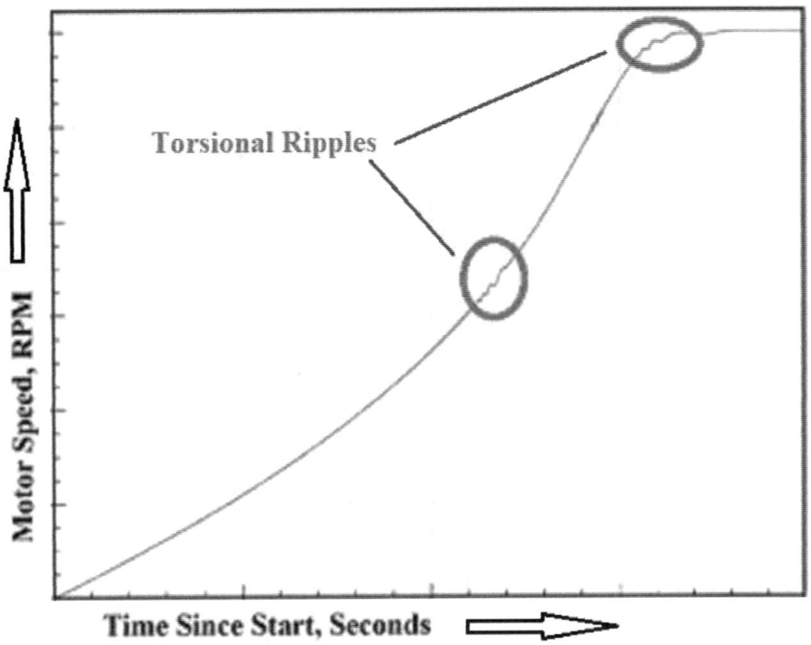

Fig 4 Torsional ripples during start up as marked in circles

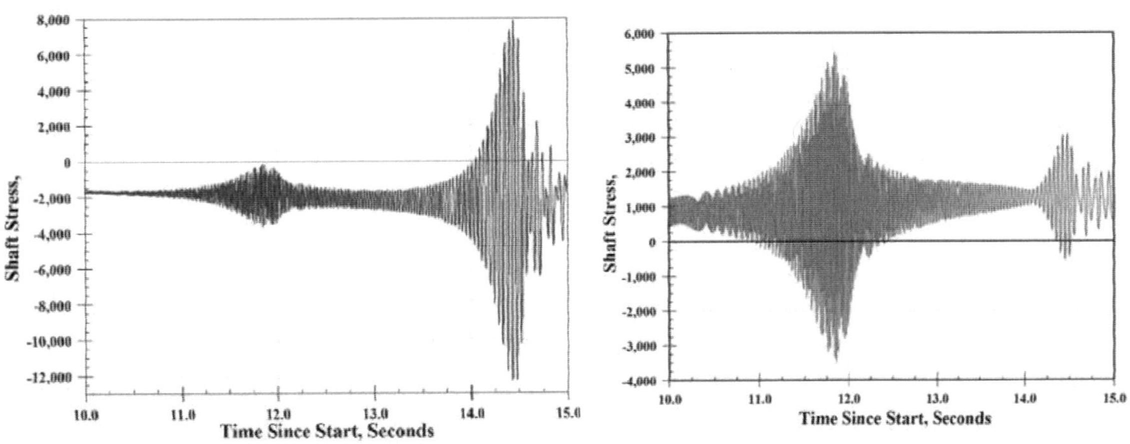

Fig 5 Stresses at Low and High-speed shaft lines during torsional ripples

(courtesy Dyrobes)

***** Formatting Issue - Best Available Paper/Graphic *****

API 617, eighth edition, section 2.8.7.5 requires that motor driven trains be designed for the start-up condition as well as steady state operation to avoid resonance with one- and two-times rotational speed(s) and electrical frequencies. Electrical fault analyses are also to be performed depending on the application requirement as per annex D of API 617. Issues related to torsional-lateral coupling are also encountered in gear box during start up causing trips on high radial vibrations.

When option II is considered as "conservative approach" as thought by end user, there is a change in Torsional natural frequencies (TNF)of lower order which are very detrimental during transient combined with higher train inertia.

In that case to mitigate the issue, the first comes as a larger spacer type coupling to practically isolate the gear box and compressor from torsional ripples during start up. The explanation is presented below in simplified form as below-

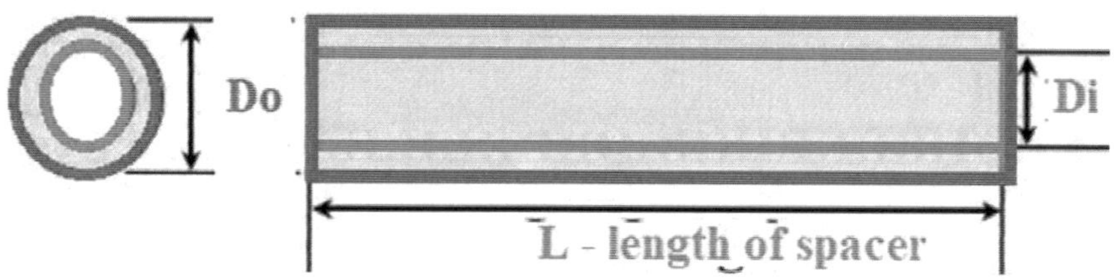

Fig 6 - Example of coupling spacer

$\Phi = TL/Ip.\ G$, $Kt = T / \Phi = Ip.\ G/L$

The above equation for a hollow pipe holds for a coupling spacer, where

Φ is the angle of twist in radians when a torque T is applied,

Ip is polar moment of inertia,

Kt is torsional stiffness and

G is shear modulus of material.

It can be seen that increasing the length of spacer L shall lower the torsional stiffness of the spacer.

Various other actions to mitigate high stresses at TNF:

1. **Soft coupling** - A coupling with rubber blocks in compression (Holset type damper coupling) generally has a lower torsional stiffness than a steel flexible disc coupling and provides additional damping. This type of coupling technically works as torsional damper. The damping limits the dynamic torque when operating near resonance. The torsional stiffness of the coupling is non-linear and sensitive to shore durometer (hardness) of the rubber blocks. Although they are a "tried and tested" option, the disadvantage of elastomeric couplings is that they are viewed as high maintenance items as the blocks need to be replaced approximately every 3-5 years.

*** **Formatting Issue - Best Available Paper/Graphic** ***

Elastomeric couplings are also temperature sensitive, dimensionally larger, heavier and produce more windage compared to conventional flexible element couplings.

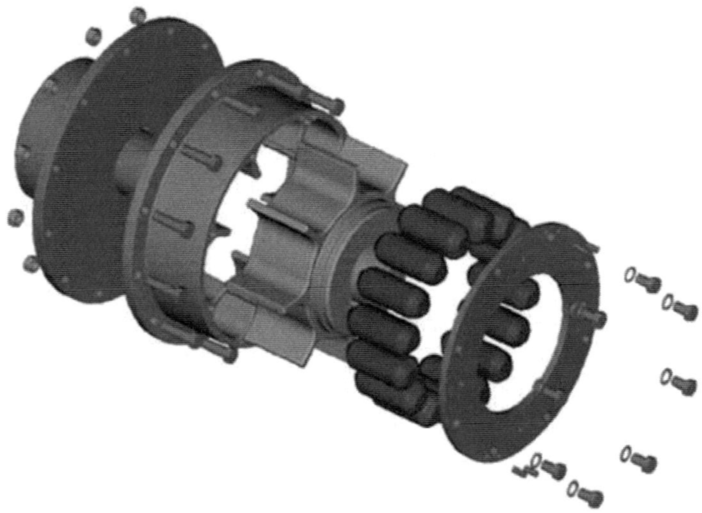

Fig 7 – "Holset" damper type soft coupling

Below table shows the comparison of soft / near soft and stiff couplings where it can be that properties are variable over ambient / working temperature. It can be that during variance of ambient temperature stiffness of soft rubber coupling varies, whereas for steel spring and disc pack coupling it remains constant. Due to change of stiffness, torsional natural frequency also varies for soft rubber type coupling.

Coupling	soft; rubber-in shear			steel-spring	stiff: disc pack
	cold	nominal	Warm		
k_C - kNm/rad	136	106	72	124	15,117
1st TNF, ω_0 - Hz	5.58	4.93	4.04	14.75	77.98

2.Hydrodynamic (Fluid) coupling:

The most practised and believed opinion is that hydraulic couplings should be treated as zero spring rate elements that effectively divide the assembly into two independent torsional systems. But in reality, zero spring rate model is only an approximation since, as, restoring forces in the fluid yield a small but finite spring rate. These restoring forces are generated by centrifugal forces in the fluid and are, therefore, proportional to the square of running speed based on the following equation for effective stiffness:

$$kh = D^5 \cdot RPM^2 / Constant$$

where Kh is Hydraulic coupling spring rate, D is Outside diameter of coupling impellers RPM is Speed of driving member (rpm).

*** **Formatting Issue - Best Available Paper/Graphic** ***

Compared to other drive elements, the boundary value of the stiffness is very low. For example, any type of coupling used the stiffness is ten times higher than that of the hydrodynamic coupling. Even stiffness of highly flexible couplings is higher by a factor of 3 at an identical nominal torque.

This coupling has the beneficial aspect of effectively dividing a torsional system into two independent systems. This permits the isolation of vulnerable components from significant sources of vibration such as synchronous motors. In addition to its isolation characteristic, the fluid coupling also provides damping to the machine by virtue of viscous friction between the impellers and coupling fluid.

C = Constant * (1+1/n^2) *HP / RPM^ 2

where: c = Damping coefficient, n -Order number of modes being excited HP -Average power transmitted through coupling, RPM -Speed of driving shaft (rpm).

Issue of concern with a hydraulic coupling is that it cannot be used with a motor having an installed power more than 6 MW with the existing design. Another issue is the higher journal loading on DE shafts due to high overhung of hydraulic coupling unit.

Coupling with controlled slip and release: This type of supplied coupling with controlled slip and automatic reset to improve performance and minimize production stoppages. They can handle dynamic torque peaks of short duration safely and without releasing. The type of coupling is completely automatic, and the self-reset function maximizes production uptime. However, as there are no refence of usage of this coupling an end user shall be very hesitant to use this.

Fig 8 - Coupling with controlled slip and release ;(courtesy Voith)

Larger coupling: A coupling spacer larger than 800 mm , if selected as to resolve TNF issues , this can cause high journal loads on Drive End side of rotor due to high share of overhung . A larger spacer may lower the lateral critical speed of coupling and hence that aspect also limit the length of coupling.

As coupling size increase, the windage losses will also increase. Due to higher windage losses, heat built up within coupling enclosure may violate the API 671 criteria (60°C), hence in order to restrict the heat built up, additional air flow / oil flow with baffle type coupling enclosure design may be required to dissipate the heat from the coupling enclosure . This will eventually lead to bigger size of lube oil system to cater oil to hydrodynamic bearing and coupling enclosure.

Requirement of material with torsional rigidity: To increase the torsional rigidity, sometimes titanium spacer used. The coupling spacer material such as titanium having more than 1000 mm length will have higher cost to the end user in terms of spares.

Other possible mitigation action and their limitation to implement:

Variable planetary gear drive:

By the use of variable planetary gear, the motor torque during start-up is approx. 18 % of the rated torque at full speed. This reduces starting torque and allows a reduced voltage start-up of the motor. This drive consists of a planetary gear and a hydrodynamic torque converter in the" power-split" branch. However, this increases the footprint of machine and adds to cost.

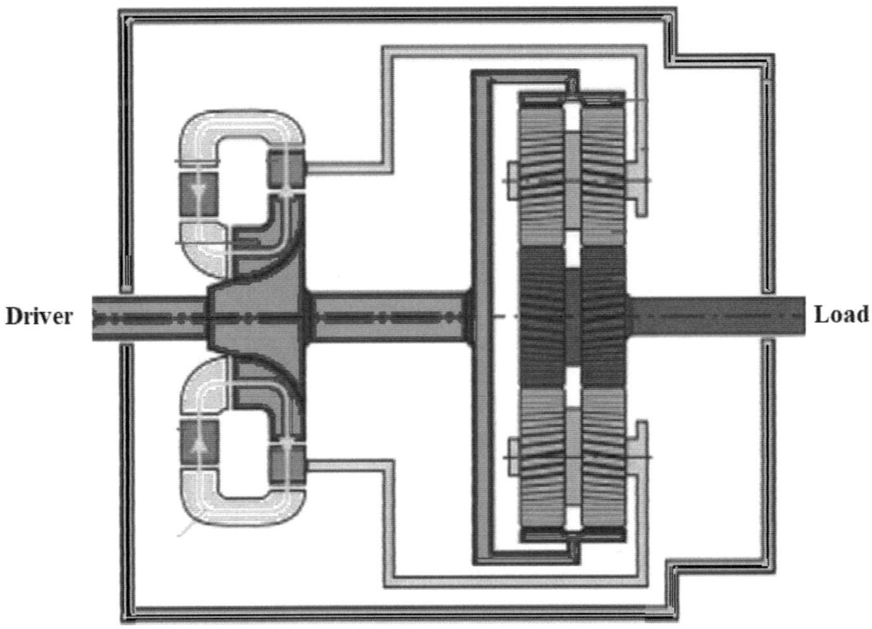

Fig 9 - Variable planetary gear drive (courtesy Voith Vorecon)

Laminated pole motors:

Laminated pole motors generate lower levels of pulsating air gap torque than solid pole motors which results in lower start-up torques and associated stress levels. However, laminated pole motors are subject to mechanical stress limitations which limit them to low speed applications.

Limitation due to Operating philosophy:

The closed compressor loop design does not allow the casing pressure to be reduced following shutdown by venting or flaring the gas due to environmental considerations. This means the compressor unit needs to be started in settle out condition, resulting in relatively higher start-up loads. The higher start uploads results to longer time to reach synchronous speed i.e. longer dwelling time in torsional ripple zone as shown in figure 4 and 5.

*** Formatting Issue - Best Available Paper/Graphic ***

Consequence of design change in torsional model due to oversized gear box:

Motor starting limitation: For synchronous motor (without variable frequency drive / soft-starter), during motor starting, high voltage drop is normally expected based on the main grid as well as higher current demand is normal. Motor starting torque is directly proportional to the voltage drop, so torque available is less (due to voltage drop) which demand higher current (4 to 5 times normal current) during starting. Heat built up within motor rotor is expressed as below with a generic formula:

$$E_{Loss} = J(2\pi n_1)^2 \int_{SStart}^{Send} S * ds \; + \; 2\pi n_1 \int_{t=0}^{t=a} S * Mm * dt$$

Where, Eloss – Energy loss during asynchronous starting (in motor rotor)

J – Moment of Inertia of driven equipment

S – Speed of driven equipment

Mm – Motor torque

n1 – No load speed,

π is constant

The time to reach synchronous speed will increase for higher (10.75 MW) rated gear box due to high train inertia requirement . This also contribute to high heat built up within motor rotor. The heat built up with in rotor pole shoe may be outside the mechanical limit and hence selected motor design may need major modification in terms of higher cooling requirement. The major modification can be bigger shaft length or may be bigger frame size with same kW rating.

Review of Option II:

With the process operational point of view, compressor worst case demand 8.8 MW of power from motor, hence gear box rating with @8.8 MW with a safety factor of 1.6 (as per API 613) can satisfy the train requirement. Since the process gas is clean and no polymerizing agents are present, chances of compressor fouling leading to higher power demand is incredible.

For 8.8 MW & 10.75 MW gear box design , actual teeth bending stress value are kept same with adjusting the loading area, which indicates that 8.8 MW gear box teeth can take same tangential load as 10.75 MW gear box, hence mechanical strength point of view 8.8 MW gear box will be adequate for the train process requirement. Refer below comparison study:

10.75MW Gear box	8.8 MW Gear box
Pinion teeth bending stress (MPa): 230	Pinion teeth bending stress (MPa): 230
Bull gear teeth bending stress (MPa): 220	Bull gear teeth bending stress (MPa): 220
Tooth pitting index (MPa):1.87	Tooth pitting index (MPa):1.87

*** Formatting Issue - Best Available Paper/Graphic ***

In order to check the mechanical suitability of the 10.75 MW & 8.8 MW gear box, tooth pitting index, bending stress calculation is performed and results are presented in below figures. It is concluded that both gear boxes have same bending stress (of tooth) with different tooth face width, which can take same amount of tangential forces.

A parametric/sensitivity analysis of both options based on possible process demand and subject compressor is carried out. It is understood that one of the main objectives of providing a driver power margin along with higher rating of gear box is to meet later-life field operations with degraded compressor performance.

One of the most cost-effective means to maintain production rates on a declining field without any external intervention is to reduce the FWHP (flowing well head pressure). However, there is a limit to how much the FWHP can be reduced. The lower limit depends on many factors, the most critical being the compression limit (pressure ratio). Pressure ratio or the polytropic head capability of any re-wheel design depends on the capacity of the existing casing to accommodate an increased impeller count or impeller diameter for a given gas composition and gear ratio (speed limit). It is important to ensure the head capacity limits of the existing casing are established,

The other objective of the re-wheel exercise was to explore whether it was possible to increase the pressure ratio capability within the existing driver power limits

Following possible operational conditions are deliberated -

1. High flow demand to utilize the motor installed power during winter season - The particular compressor performance curve shows a rapid drop in performance; mainly based on type of impeller selected (impeller exit angle) or to meet rate of rise of polytropic head requirement by end user. Refer below figures for compressor performance curve and effect of impeller exit angle. Based on fan laws, can be understood that installed compressor with fixed speed driver cannot meet the target discharge pressure if mass flow is increased. From the compressor performance curve, even 10% increase in flow will have sharp fall in polytropic head and the unit cannot meet the gas transmission manifold pressure.

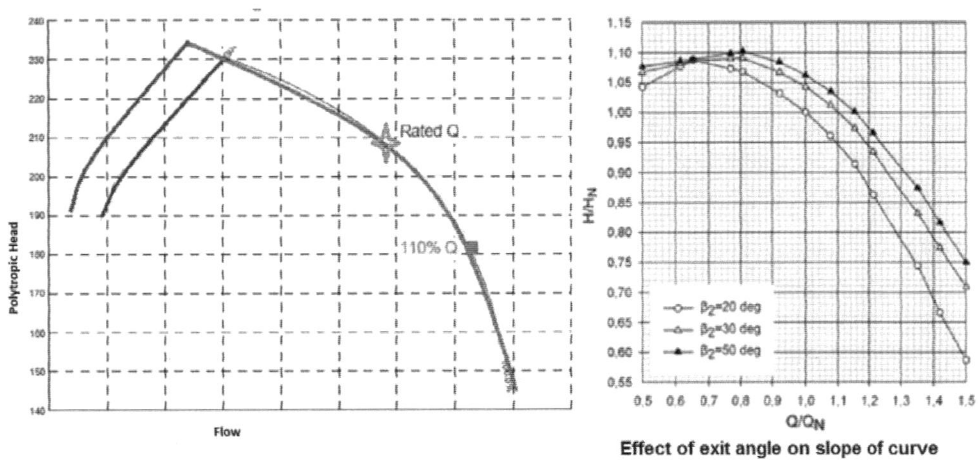

Fig. 10- Compressor performance curve and effect of impeller exit angle on slope

***** Formatting Issue - Best Available Paper/Graphic *****

2. Lower suction pressure or higher discharge pressure - In case of depleting gas well with which the compressor suction train is connected, low suction pressure may be a possible operating scenario. To meet a higher target polytrophic head, compressor wheel peripheral speed needs to be increased. Without changing the compressor impeller this cannot be achieved with fixed speed synchronous motor drive.

Further to consequence of increase in pressure ratio and polytropic head, discharge temperature may increase rapidly and may go out of boundary of temperature limitation of seals, design temperature of casing and other internals. Even changing the impeller to higher diameter shall not render any help with a straight through design as no inter-stage cooler can be introduced. It is to be noted that a larger impeller causes a larger heat to rise due to disk and skin friction.

Table - Sensitivity analysis with depleted well condition :

k (Cp/Cv)		1.3125	
mol wt		19.4000	
suction pressure		18	barA
suction temp in degC		48.9	
suction temp		322.05	kelvin
actual suction temp		337.671	
discharge pressure		79	barA
mass flow rate		97200	kg/hr
mass flow rate		27.0000	kg/sec
Comp factor Z1		0.96	
etap*		0.8	
R		**8314.34**	
head / stage		3281.265771	meters
Select Wheel diameter		400	
Number of stages		7.298159109	
Iterative Round off		7	
Ad .work factor	0.55-0.66	0.6	
		61018.70879	
Tip velocity		247.0196526	m/sec
N RPM		11800.30188	
Stage eff. factor		0.8	
Gas Horse Power		**7.929**	MW
Installed power		**8,895.910**	KW

k (Cp/Cv)		1.3125	
mol wt		19.4000	
suction pressure		15	barA
suction temp in degC		48.9	
suction temp		322.05	kelvin
actual suction temp		337.671	
discharge pressure		79	barA
mass flow rate		97200	kg/hr
mass flow rate		27.0000	kg/sec
Comp factor Z1		0.96	
etap*		0.8	
R		**8314.34**	
head / stage		3281.265771	meters
Select Wheel diameter		400	
Number of stages		8.38897184	
Iterative Round off		8	
Ad .work factor	0.55-0.66	0.6	
		61371.4629	
Tip velocity		247.732644	m/sec
N RPM		11834.36197	
Stage eff. factor		0.8	
Gas Horse Power		**9.114**	MW
Installed power		**10,225.529**	KW

Remark / Observation	**Limitations:** Discharge temperature out of limit.		**Limitations:** Discharge temperature out of limit	**Limitations:** 1. If mass flow increased from rated flow by 10%, reduce the efficiency sharply & polytrophic head requirement does not meet. *** 2. Flow co-efficient increase on border line to 0.1.

*** Formatting Issue - Best Available Paper/Graphic ***

The non-dimensional operating points in terms of flow coefficient, head coefficient combined with work coefficient of subject compressor considered, they appeared to be traverse out of boundary of acceptable region.

*Even if re-bundling is decided with same casing and fixed speed synchronous motor, polytrophic efficiency improvement may demand new impellers with higher axial lengths. Improved polytrophic efficiency may lower power requirement at gear box shaft end however casing length may not be suitable to cater increased rotor length due to higher axial length of 3D impellers.

**Inter-stage cooling shall be required which needs a 4-nozzle casing. (refer fig 11)

*** If impellers are to be changed with larger diameter for higher peripheral velocity to meet required polytropic head , return channel , vaned or vaneless diffuser and casing frame may limit re-bundling with the required bigger impellers as bigger impellers may lead to larger cross over channels .

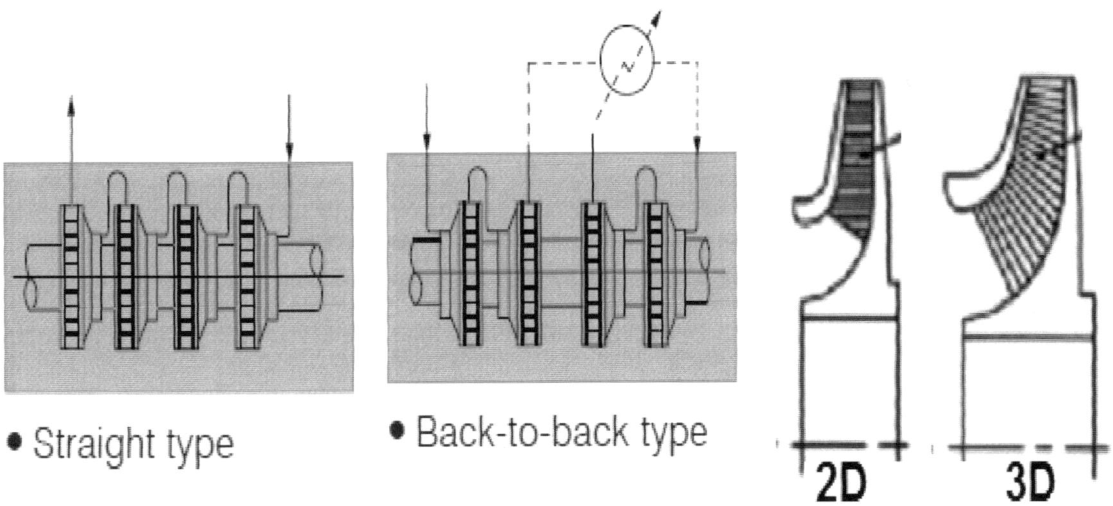

Fig 11 - Impeller configuration and axial length of 2D and 3D impellers

Observations &Conclusion:

- There is a high probability of torsional instabilities during motor start up, with a longer duration, if load (gear box bull gear) inertia is high. Particularly for synchronous motors higher torsional oscillations during start up is always a potential issue.

*** Formatting Issue - Best Available Paper/Graphic ***

- To tackle the issue of high torsional oscillations is complex and costly mitigation actions are required with a domino effect of higher cost of operation.
- Process parameters in terms of depleting wells (design for future), configuration of impellers in compressor casing and Slope of Compressor performance curve have a major role on plant operation.

It can be concluded with above dummy case study that following the bullet point of API 613 5th edition clause 2.2.1, rating a gear box just based on electric motor nameplate rating is not a prudent idea to utilize full motor power.

Before enforcing API 613 bullet point clause to a project specification, a detailed analysis of futuristic operation profile can help to decide the compressor impeller arrangement when the gear box rating clause is applied.

In case the re bundling is required, it is possible to explore high pressure ratio type compressor which can match the design and speed ratio of gear box. This type of compressors has high polytropic efficiency and need lesser stages than standard design.

If the operating case is defined, then OEM do not need to apply the bullet clause; the gear box shall be selected based on maximum brake power demand from compressor. The same reasoning should be discussed and deliberated with engineering team of end user and EPC contractor.

If operating cases are not defined and includes future cases with depleted wells, then OEM may need to apply the bullet clause, i.e. the considering motor name plate rating and safety factor for the re-bundled configuration of the compressor. The same scenario should be discussed and deliberated with engineering team of end user and EPC contractor

References:

1. API 613, 5 Edition, "Special Purpose Gear Units for Petroleum, Chemical, and Gas Industry Services", American Petroleum Institute.
2. A Comparison of Current AGMA, ISO and API Gear Rating Methods - John M. Rinaldo [GEAR TECHNOLOGY | July 2018]
3. Dudley's Handbook of Practical Gear Design and Manufacture, Second Edition
4. Fundamental Rating Factors and Calculation Methods for Involute Spur and Helical Gear Teeth- ANSI/AGMA 2001- D04
5. Practical implementation of torsional analysis and field measurement - Malcolm E. Leader, and Ray D. Kelm, -Source – Dyrobes
6. Start-up Torsional Vibration of Rotating Machine Driven by Synchronous Motor- Takuzo Iwatsubo, YoshitoYamamoto &RyojiKawai - Dynamics of Multibody Systems IUTAM/ IFToMM Symposium, Udine, Italy, September 16–20, 1985
7. Torsional Vibration Analysis and Testing of Synchronous Motor-Driven Turbomachinery. - Mark Corbo et al, 31st Turbomachinery symposium Texas A&M
8. Revamping A Gas Compressor Drive Train From 7000 HP to 8000 HP With A New Synchronous Motor Driver and A Controlled Slip Clutch Mechanism, - V. Hutten, et al 35th Turbomachinery symposium Texas A&M

*** Formatting Issue - Best Available Paper/Graphic ***

9. Turbomachinery Symposium, Turbomachinery Laboratory, Texas A&M University, Texas
10. Torsional Vibrations – a (twisted) Overview, Notes 9, Dr. Luis San Andres Texas A&M
11. The Application of Flexible Couplings for Turbomachinery – Mancuso et al; Proceedings of 18th Turbomachinery symposium Texas A&M
12. Lateral Critical and Train Torsional Analysis and Rotor Balancing, API Publication 684
13. Compressor Performance: Aerodynamics for the User, 3rd edition - M. Theodor Gresh, Elsevier
14. Optimizing Component Selection in Synchronous Motor Compressor Trains Based on Technical and Financial Considerations Martin D. Maier et al -46th Turbomachinery symposium Texas A&M
15. ISO 6336-3:2019(en) Calculation of load capacity of spur and helical gears — Part 3: Calculation of tooth bending strength
16. http://voith.com/corp-en/torque-limiting-couplings.html?102340%5B%5D=3

Acknowledgements:

Authors wish to tender their sincere thanks to MFPT chairman and board of directors for all their supports.

CASES OF HIGH SPEED PINION VIBRATIONS AT LOW LOAD CONDITION IN API 613 TURBO GEARS

Mantosh Bhattacharya

Petrofac International UAE

Mantosh.b@petrofac.com

Abstract:

During operation of turbomachinery train used in oil & gas , petrochemical and refinery industries the selected gear box models are expected to show up vibrations within acceptable limits as mandated by API 613 standard or jointly agreed by OEM (Original equipment manufacturer) and End user for on-site operations. Based on these agreed values, reliability team sets up pre- alarm, alarm, and trip values. However, during low load operation during mechanical run test / coast down of a turbomachinery train, the pinion of speed increasing gear box may show up high vibration and cause concerns to end user.

The proposed paper discusses several case studies where high vibration issues in turbo gear pinion were faced during various scenario such as API 613 mandated No / low load full speed run / during coast down or operating the train below actual speed map at site. With deep delving into the matter , the issues concluded as benign and non-intrusive to machinery integrity.

The paper concludes with a point for requirement of deeper understanding to address such vibration phenomena during no load / coast down to decide whether issue to be considered as genuine or benign.

Key words – sub synchronous vibration: casing resonance: torsional-lateral interaction: whirl: axial shuttling

Introduction:

Some cases studies are presented under this topic where purely lateral and torsional -lateral coupled vibration were observed which were concluded as benign.

Excitation of an unloaded pinion with sub-synchronous vibration :

During unloaded gear box spin test as per API 613 mandate , high non-synchronous (sub - synchronous in nature)vibration observed at high speed pinion shaft as shown in fig 1 . The synchronous component was lower but no synchronous components caused vibration beyond acceptable limit (shown in blue) . Further investigation revealed a sub-synchronous component ant much lower multiple that 0.4 x was present . Several times, it has been observed that as due to high weight ratio of Bull gear and Pinion shaft, the pinion shaft shows up with higher overall vibration than API acceptable criteria of sub-synchronous vibration.

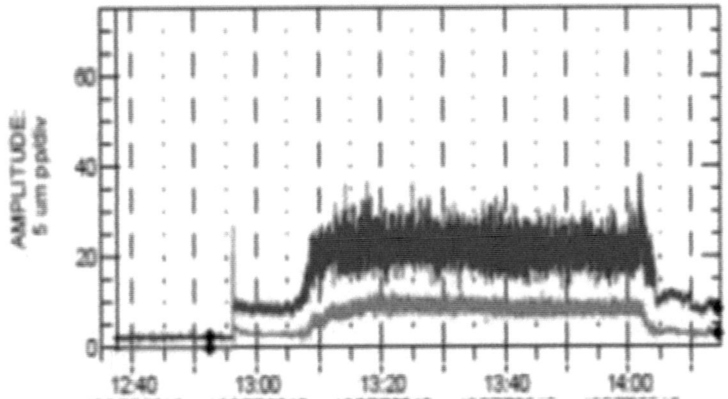

Fig 1 – Overall high vibration than API 613 criteria due to sub-synchronous component

This in fact normally occurs due to unloaded pad flutter, starvation effect or instability due to lack of effective dynamic stiffness.

Fig 2 – Relief created in leading edge of pinion bearing

During no load or part load condition, the load on the bearing, generated by torque transmission via the upward gear mesh, works against the gravitational forces of the shaft. This may generate more unstable condition in respect of rotor-dynamics, compared to a specific load applied at driven end of pinion. This situation becomes more complicated with higher the gear ratio, i.e. the higher the weight difference between pinion and bull gear shaft. If this issue occurs at test bed , considerable time is wasted in analysis at OEM works test bed and engineers rush to incorporate lead edge modification .Leading edge geometry of pads of pinion bearings are normally modified by scraping babbitt (shown in red markup)material to increase the oil supply. (as shown in fig 2) or creating a dam at lower cylindrical portion of bearing. In fact, this concern is based on mandate from API 613 as - If the amplitude of any discrete, nonsynchronous vibration exceeds 20% of the pinion shaft (or 40% for shafts running below 2000 rpm) of the allowable vibration as defined in 2.6.6.5, the purchaser and the vendor shall mutually agree on requirements for any additional testing and on the gear unit's suitability for shipment.(cl. 4.3.2.2.9).

It was observed that in some of site installation that non-synchronous vibrations with high amplitudes were encountered on the high -speed components (pinion shaft)of the train found to be excited by the heavy input shaft assembly (gear wheel / bull gear) of gearbox. The mechanism of excitation was due to 'backward whirl' motion of the bull gear shaft if equipped with the sleeve bearings. These vibrations were found to be sensitive to oil inlet temperature (lower amplitudes with lower oil temperature and vice versa). This had led to the design of new gearbox output shaft sleeve bearings with higher damping characteristics. This activity caused delay in startup of machine. It is expected the designed bearings should suppress the nonsynchronous vibrations below the acceptable limit.

However , the issue was deftly tackled by incorporating minor tweaks (increasing oil supply pressure and flow a little) . The overall and sub-synchronous vibrations came down and no load test was passed as shown in fig 3 .

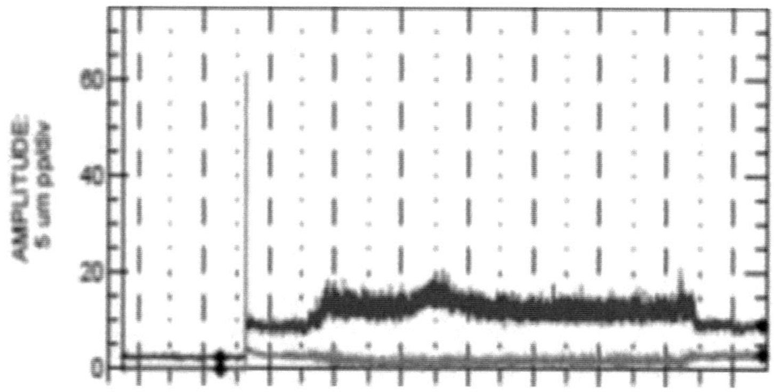

Fig 3 – Overall vibration came down below 25 um pk-pk after minor adjustment

However two sets of bearings were kept ready for possible modifications on leading edge if sub-synchronous vibration showed up during full load full speed complete unit test of turbo-compressor train . No sub-synchronous vibrations observed were during full load full speed complete unit test .

To address all such probability, non-synchronous pad coefficients may be considered as well during rotor-dynamic study during design phase if the bearings selected are new to load and speed combination. It is always better to advise OEM to provide reference list where they applied pressure dam / recess modification on supplied API gear units when this type of issues is encountered.

It is advisable to test gear boxes with at least 70 % of load or utilize the full load full speed complete unit test where it can be assured that units shall not face any sub-synchronous vibration during full range of operation window at site . Actual test of rotor-dynamic integrity of gear box is with load so OEM should be proactive to get the alternative design ready with quickest support from bearing manufacturer.

Effect of shop motor rating and influence of couplings in Mechanical run test: High vibrations were observed in bull gear of turbo gear during API 613 mandated full speed no load Mechanical run test. This turbo gear was part of electric motor driven centrifugal compressor train. The FFT analysis revealed high vibration in bull gear at sub synchronous frequency, test run was called off as any possible cause could not be detected. However, the trend plot detected a wave like pattern on the X-probes for both end of the bull gear bearings. It was varying between 21 microns to 15 microns for (Non drive end) and 16 to 8 for the (Coupling end). As jointly discussed with OEM specialist and test bed engineers, the test was started again after pre-tension of the coupling and installation of the Z probes.

It was found a cardan shaft coupling used in MRT with a high rated electric motor (much more than actual power required to conduct the test). It can be seen in trend plots that a wave like pattern appears on the X-probes for both end of the bull gear bearings. A low frequency vibration was found at same frequency of axial shuttling as shown in fig below –

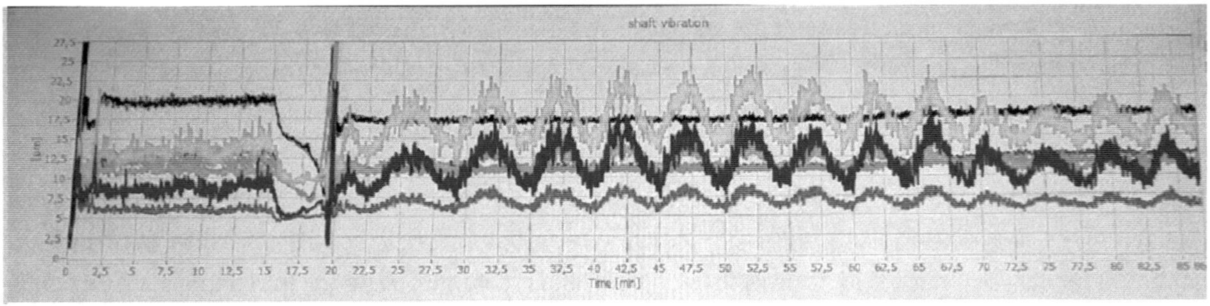

Fig 4– Wave like modulated Bull gear vibration plot

The coupling was replaced with a dry flexible coupling. The test was started again after pre-tension of the coupling and installation of the Z probes (to detect. Low frequency (non-synchronous) vibration can be seen in below FFT plot as well and wave like pattern remained.

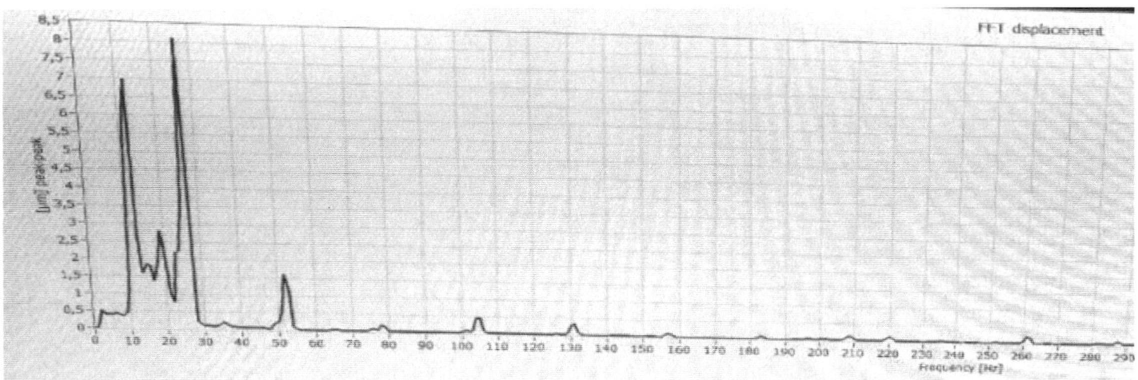

Fig 5 -Non-synchronous low frequency peak at full speed test

It was investigated further and explained that it is a phenomenon occurring because of the low load required (270 KW) during the testing. The hunting of magnetic center of motor and axial force exerted on coupling caused a modulation effect on bull gear vibration. It was explained that the motor is exerting axial force and due to low load, the gear is displaced slightly and resulting in the wave like pattern. As the overall vibration levels are with in acceptance range, the vibration level in sub synchronous frequency was low during the final run meeting API criteria. To verify this explanation, the axial probes were installed for the test running of gear box with special tag numbers and connected to shop vibration monitoring system r. It was found that the pattern in axial probe is also same as vibration on probes on bull gear. It was explained by specialists of gear box OEM and compressor / motor OEM that the shop motor is exerting axial force and due to low load, the gear is displaced slightly and resulting in the wave like pattern. As the overall vibration levels are with in acceptance range, it is normal.
To verify this explanation, the axial probes were installed for the test running of another turbo gear unit of same batch which were ordered for all four turbo-compressor trains. It was found that the pattern in axial probe is also same as vibration on X probes on bull gear.
It was requested to OEM for replace the test motor with lower KW motor. OEM specialist also agreed that this phenomenon is because of low power (Power required 270 kw as against installed 3600 KW) during the testing. There is axial force exerted by the motor, which is causing the gear to displace, and it is not because internal gear issue. It was clear to see that this phenomenon is caused by the Motor and not generated from the gear box but compounded by gear type coupling / cardan shaft used in test bed.
Motor mag center hunting is common in sleeve bearing machines. The motor's centering force can be about 100 lbs. per 1000 hp as a rule of thumb. This centering force may cause the motor rotor to vibrate axially like a simple spring mass system. The frequency could be low and shall always affect the bull gear- pinion mesh if flexible non lubricated coupling is not used at no load test. Geared coupling aggravates this issue.

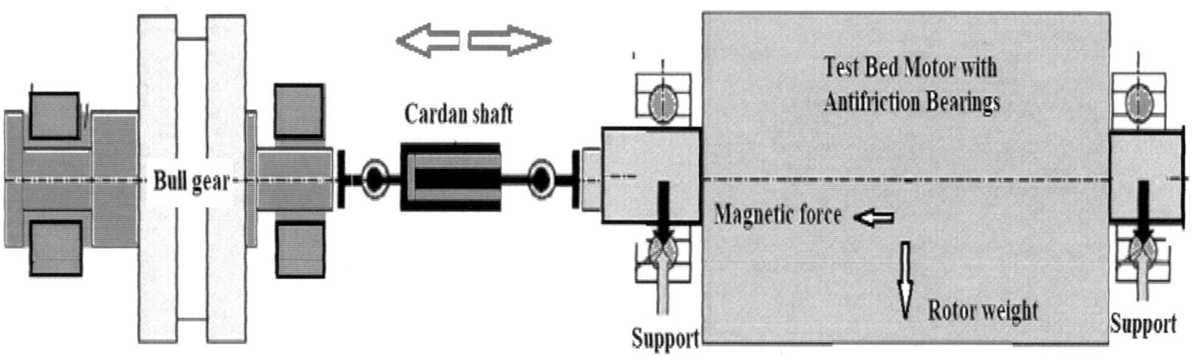

Fig 6 Axial shuttling due to magnetic force from electric motor to unloaded gear box

A low powered motor (fit for purpose for No load test) with antifriction bearings shall exert less centering force and shall lower down modulations (shuttling) as seen in spectrum in previous tests. With new test motor, as predicted, the wave pattern is subdued as below

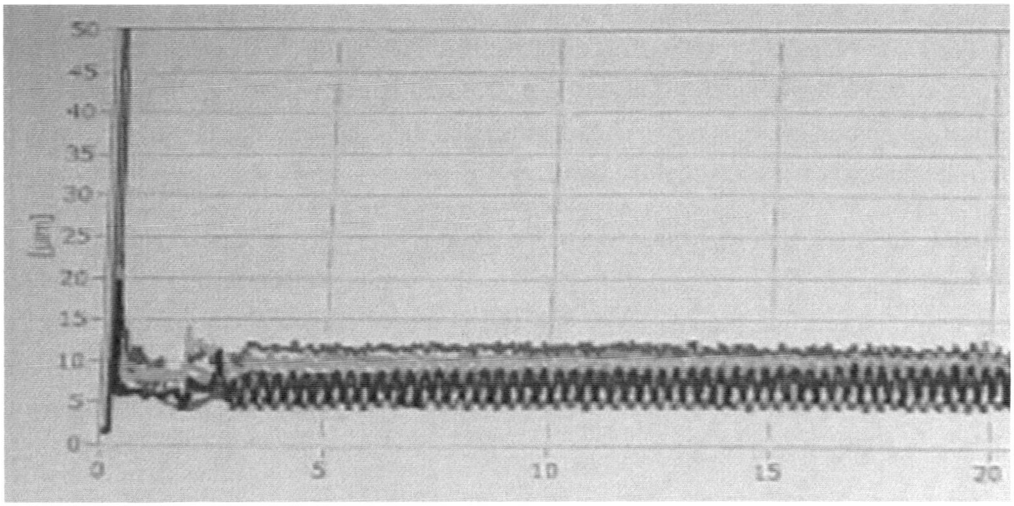

Fig 7 – Wave lime trend with much lower amplitude

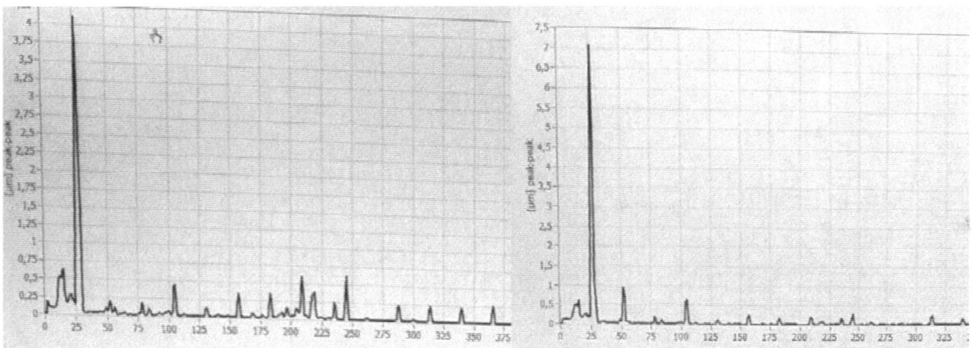

Fig 8 - Low frequency component lowered

The point was deliberated into all possible failure modes considering as technical glitch and jointly understood that job motor is equipped with slide bearing and shall always be operated with larger inertia and load , this phenomena shall not occur again at site . This phenomenon will subside when gear is put on load and because of gear load this axial force will become negligible as explained and understood. The compressor unit was commissioned, and no such issue was reported again.

Excitation of torsional- lateral coupling on turbo-compressor train:

Subject machine - VFD driven 12 MW electric motor turbo-compressor with double helical speed increaser gear box. A high vibration peak reaching above alarm level occurred on the high-speed pinion shaft of the gearbox during coast down generated by pushing normal stop button. While the complete train was coasting down the pinion shaft vibration peak exceeded the trip settings of the installed machinery protection system. It raised the expected concern to end user. It was observed that no high vibration was observed in any component during startup of train. Before attempting to restart the machine, it was decided to check all historical data of pinion vibration during previous start up and coast down.

It was observed that some of previous shutdown did not cause any high vibration in pinion. After detailed study it was found that shutdowns were due to a spurious trip signal caused by dry gas seal system or by level 3 process trip . Those trips took the course of fast shutdown as per control logic. No high vibration was detected during coast down.

It was further found that it took 7 mins 30 sec for the high-speed shaft speed to fall from 5000 rpm to @832 rpm (when the high vibration occurred) when a "NORMAL STOP" was executed. In case of previous shutdown such as EMERGENCY SHUTDOWN'S it took less than 1 minute for the shaft speed to fall from approx. 5000rpm to approx. 832 rpm. This means the shaft passes through the vibration peak much faster, so it does not have so much time to build up to such a high value. See the attached trends to compare.

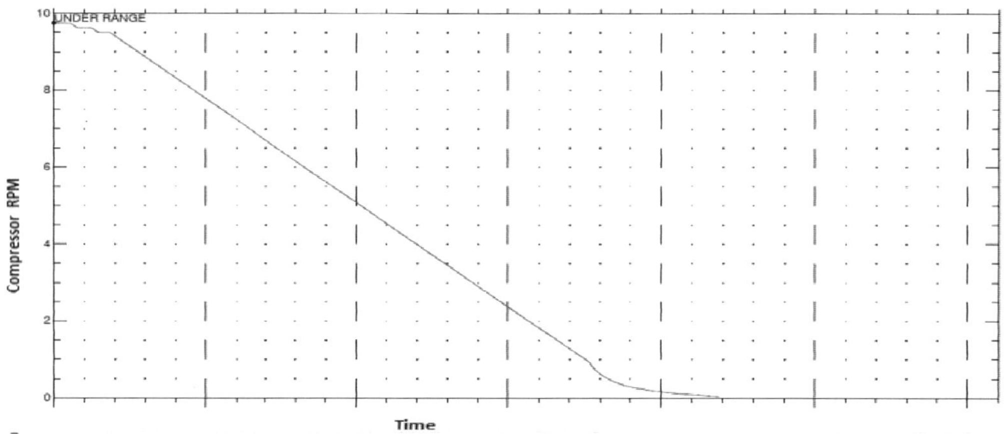

Fig 9 – Normal shutdown -speed vs time profile

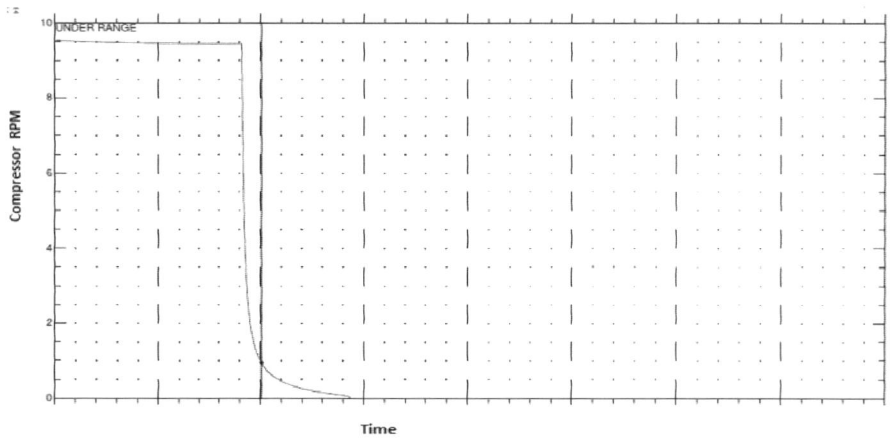

Fig 10 –Fast stop shutdown -speed vs time profile

The maximum displacement of the high-speed shaft was still well within the clearance of the bearing. Further investigation revealed the high vibration was due to coupling of torsional mode into lateral vibration mode which excited the bull gear.

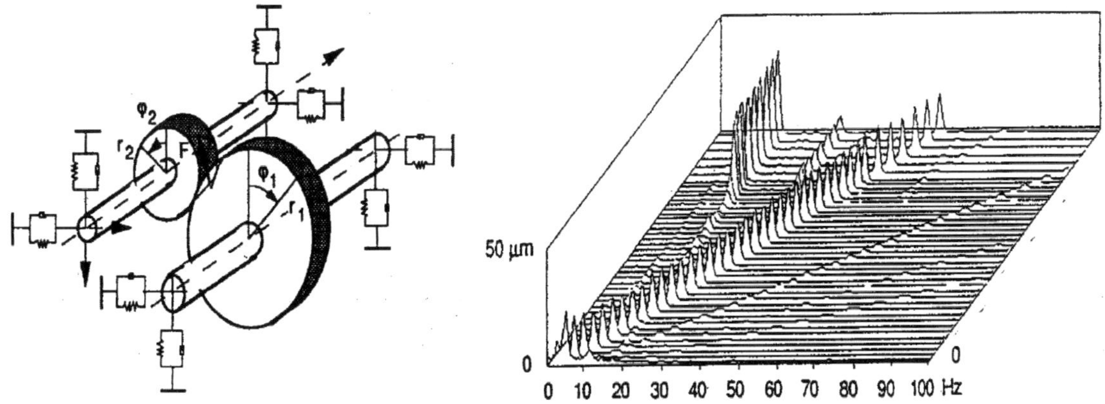

Fig 11; Torsional -lateral coupling effect on pinion with high vibration

As the bull gear had a higher mass, it excited the pinion in turn. During the normal stop, the train dwelled into a TNF 1st mode zone 825 rpm (+/- 10% range) of and it had sufficient time to create a spike of high lateral vibration.

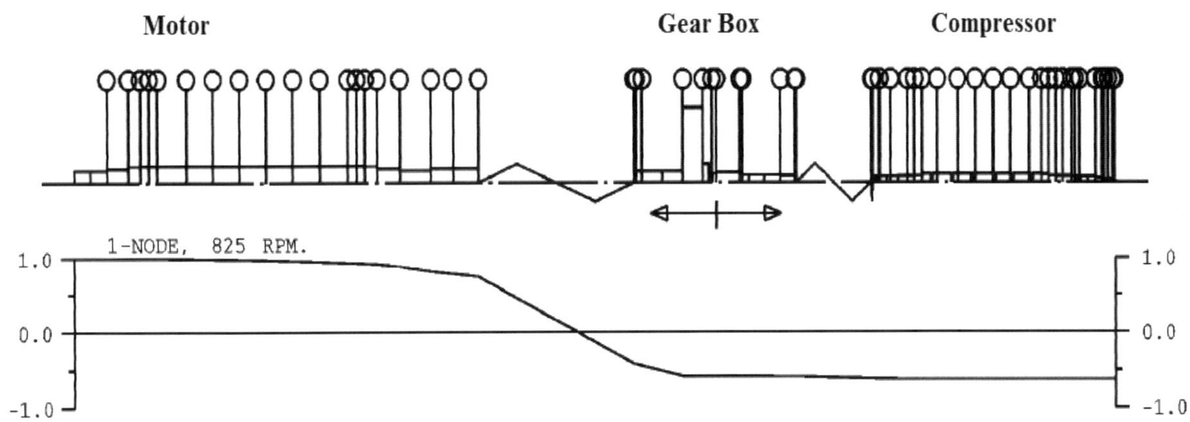

Fig 12 – Relevant Torsional mode shape of compressor train

. After further review of the data it was found that VFD motor is cutting out when the speed of the motor / gearbox low speed shaft is @ 125rpm and the compressor / gearbox high speed shaft is approx. 834 rpm. This was causing the axial shift in the gearbox and the pinion to respond with the high displacement at that point. The issue was closed based on low energy of vibration during coast down and sufficient clearance between journal and bearing.

Another case of excitation of pinion due torsional- lateral coupling during unit test
During full load complete unit test of VFD motor driven centrifugal compressor train at OEM test bed, it was observed that vibration of gearbox pinion (high speed side) was exceeding the

permissible limit at lube oil supply temperature of 56 deg C. This limited the testing of compressor at 65 deg C of oil temperature as envisaged in the test procedure.

Vibration at 56 deg C reached up to 32 microns in few readings when 2.5MW load on compressor was there. Following observation were made at OEM test bed in following order –

•The train was running with speed increasing gear box far below the API vibration limits (25 um pk-pk) until the oil inlet temperature was raised above 57°C.

•The vibrations of the gear box high speed pinion shaft at 60°C reached a value of >40 μm. No changes in the compressor rotor vibration were observed.

•The vibration increase identified discrete response peak at 65 Hz.

•Variation of oil supply pressure and volume flow resulted in only small changes of the vibration level.

•The compressor load of 2,5 MW was increased to 3,8 MW.

•Three readings with constant oil supply temperature of 65°C and speed and load variations were taken:

•65°C 105% speed 3,8 MW load ~10 μm p-p

•65°C 105% speed 3,7 MW load ~48 μm p-p

•65°C 80% speed 1,44 MW load <10 μm p-p

The observed vibration peak at 65 Hz is very close to the calculated second torsional natural frequency as 63 Hz.

•It was observed that under certain combinations of oil supply temperature, speed and load the damping of the high-speed pinion bearings decreased.

When OEM engineering team was contacted, it was anticipated this is being excited from the input of VFD affecting the second mode of torsional critical speed.

Witnessing team and test bench engineers did the trial running of compressor by increasing the load from 2.5 MW to 3.5 MW load. Gear box pinion vibration has come down to 7- 8 micron.

•From the measured points a load versus speed restriction (no dwell zone) for oil supply temperatures >57°C was be derived.
•All specified compressor operating points are well outside the no dwell zone.
•Only a very small area of the test case with nitrogen performance map is affected by this restriction.

•Torsional excitations due to VFD are usually well damped and do not result in high lateral vibrations even if close to torsional natural frequencies.

•Under certain combinations of oil supply temperature, speed and load the damping of the high-speed pinion bearings decreases.
•Load limit according to the diagram shown will be implemented.
•No specified operating point is affected by this load limit.
Train HABGAS 1 is operationally safe within the specified oil supply temperature, speed, and load ranges. It was jointly agreed with full deliberation and prudence that No modification in existing train is needed and unit was towed out from OEM works.

Whip like phenomena during compressor train coast down -

During a process operation related stop of gas turbine driven centrifugal compressor, high vibration amplitude was observed during coast down on the one of gearbox pinion bearings crossing the HH alarm and reaching an amplitude of 180 um pk-pk. This caused a major concern in operation team to deliberate whether the open the gear box after calling OEM engineering representative as the equipment was in warranty period.

Following points were deliberated:

The gearing forces lift the pinion (up mesh type combination) then we may have a bearing instability at a speed range where the meshing force, oil wedge, and gravity cause the pinion to become unloaded. Gear bearing journals have a lot of surface area because of gear forces, radial thermal growth, etc. and without load can become unstable.

. Unless some form of bearing is utilized that can inhibit whirl it is likely that it goes into whirl as it passes through the center of the bearing clearance. It was also verified to examine the startup and shutdown shaft centerline plot and no instabilities and encroachment were found which could have damaged the bearing white metal

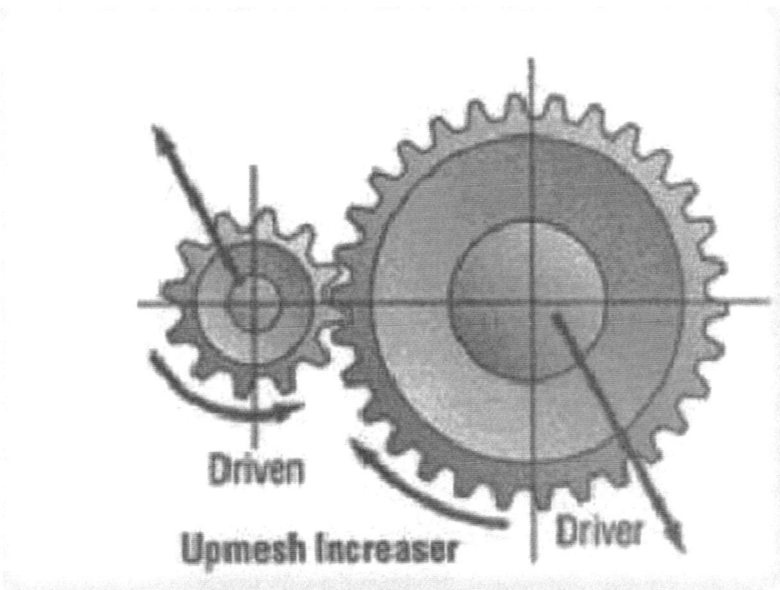

Fig 13- Illustration of Meshing direction of a speed increasing gear box.

Following aspects also checked related to anti whirl design. If the bearings are anti-whirl design but are installed backwards, they will whirl. Lemon bore (elliptical) or offset sleeve (to achieve an elliptical arrangement) bearings do not eliminate vibration instabilities if the load angle lies in the major axis of the ellipse, since the oil film stiffness in this region may not be sufficient to prevent vibration instabilities. If bearing instabilities are experienced, a possible modification is to rotate the major axis of the ellipse to provide sufficient oil film stiffness at the load angle. This procedure takes time and must be repeated in the field each time the machine is disassembled. If the above procedure is not successful, installation of multi-lobe or tilting pad bearings will be required. In this case tilting pad bearings are provided so this possibility was eliminated.

Upon diagnosis of the event through vibration plots a high sub-synchronous frequency with forward precession was observed during the coast down. Orbit shape on the GB bearings which is nearly circular, the rotation of key phasor dots is more jumbled, indicative of possible oil whip. In case of oil whirl, the dots will be migrating very slowly against rotation. The waveform also shows one vibration cycle in more than 2 revolutions, i.e. <1/2X. Average shaft centerline plot was showing a low eccentricity. This phenomenon indicates a combined presence of fluid whirl and whip which is a symptom of fluid oil instability.

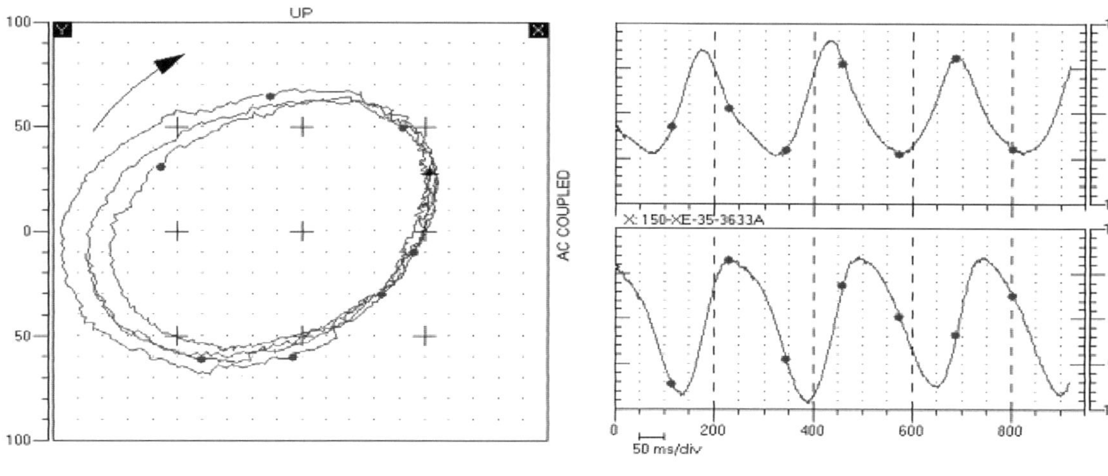

Fig 14: “Whirl like” orbit and wave form at 600 RPM of pinion

Short note on rotor instability as whip – The fundamental understanding is - oil whip occurs on those machines which are subjected to oil whirl. When the sub-synchronous oil whirl frequency (0.43 -0.47 times) coincides with and becomes locked into a system’s natural frequency (often with operating frequency or critical speed frequency). Assume the rotor speed increased to just above a certain RPM with a sub-synchronous oil whirl, its speed reaches to 2 times its first balance natural frequency. At this time the oil whirl, which was approximately 43 - 47 percent of RPM, was brought into coincidence with this critical speed. The oil whirl was suddenly replaced by oil whip - a lateral forward precession sub-harmonic vibration of the rotor.

Dry whip occurs in journal bearing machines subjected to either a lack of lubrication or the use of the wrong lubricant. When this occurs, excessive friction is generated between the stationary bearing and rotating journal. This friction can excite vibration in the bearing and other components.

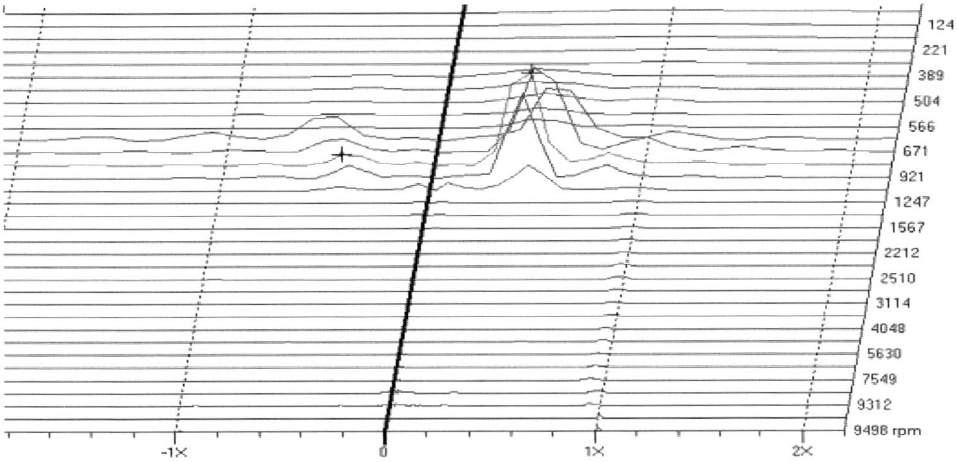

Fig 15-Water fall plot showing high vibration at sub-synchronous order

Following were the observation upon analyzing the various plots –

- The incident occurred at fast stop and with just 30 seconds duration at very low speed which was nowhere near to any critical speed or system natural frequency.

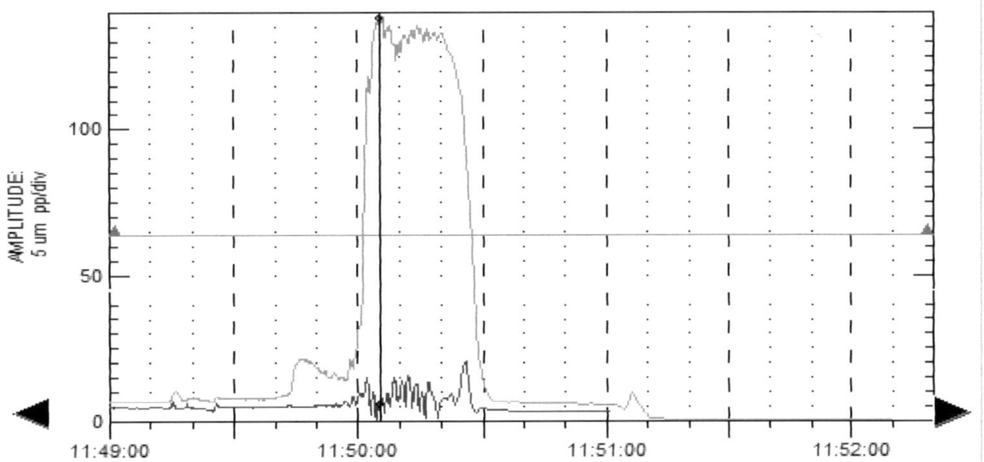

Fig 16 -Duration of whirl like phenomena during fast stop

- During that short period of time, pinion rotor was very close to being at center of bearing (low eccentricity) and caused a self-excited "whip like phenomena" due to higher clearance all over causing a temporary instability.
- The amount of circular orbit of pinion in um pk-pk was quite less than minimum inbuilt running clearance (180 μm vs 290 μm).

As bearing was unloaded during coast down and whirling the gear was further reviewed for possible thermal aspects such as growth and gear box casing temperature and offset . After repeated simulation of each train to identify the repeatability of such phenomenon , it was considered as non-intrusive to machinery integrity and the compressor trains were handed over of commercial production and gas pumping.

Cases of Gear Box casing resonant condition –

In a remote site, during commissioning of motor driven turbo compressor train end user had been observing high casing vibration at gear box of both the motor driven turbo compressor train during start up. It used to reach to alarm level and peculiar sound also observed during that time. This was observed during a certain speed range and later used to subside, and vibration level used to come back in normal. Upon reviewing the velocity / acceleration spectra, all peaks for normal frequencies which are present in a normal gear box found to be acceptable. No unusual high frequency vibration observed. Operating team could not decide on any resolution.

After detailed review of time and speed base plots of various start up events, it was found that gear box casing vibration is sensitive to a speed band only, it was not affected by any load change in compressor.

The time-based plot shows that in a band of 1100 RPM of motor / 9500 – 9600 RPM of compressor, gear box used to show up high frequency vibration.

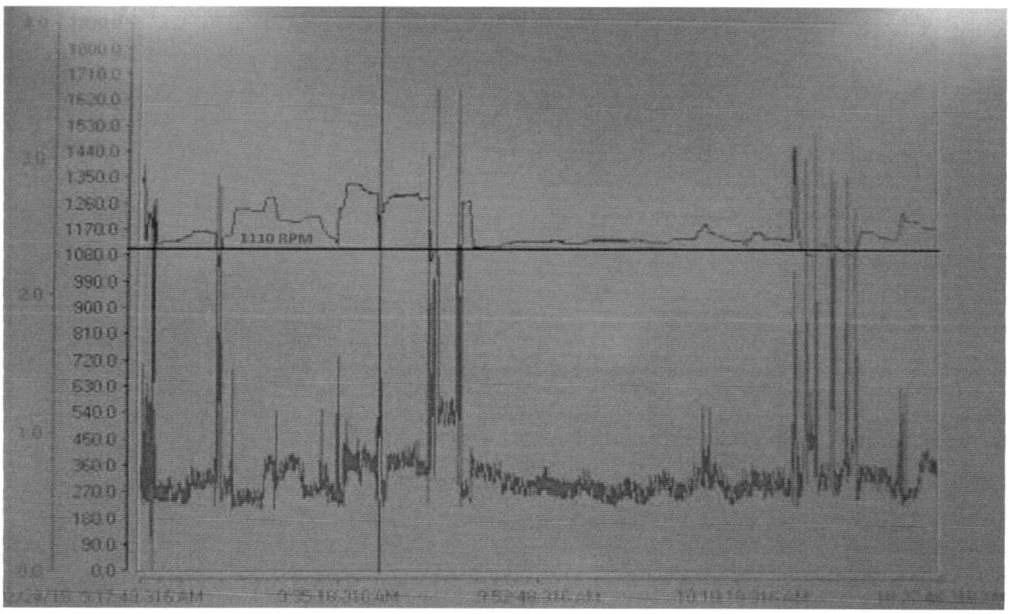

Fig 17 - Gear box casing high vibration at 1100 RPM of motor speed

It was decided to further investigate with a water fall plot keeping the speed 9600 RPM constant to detect which forcing frequency is causing this effect and it was found that a frequency of very high order was present as shown below –

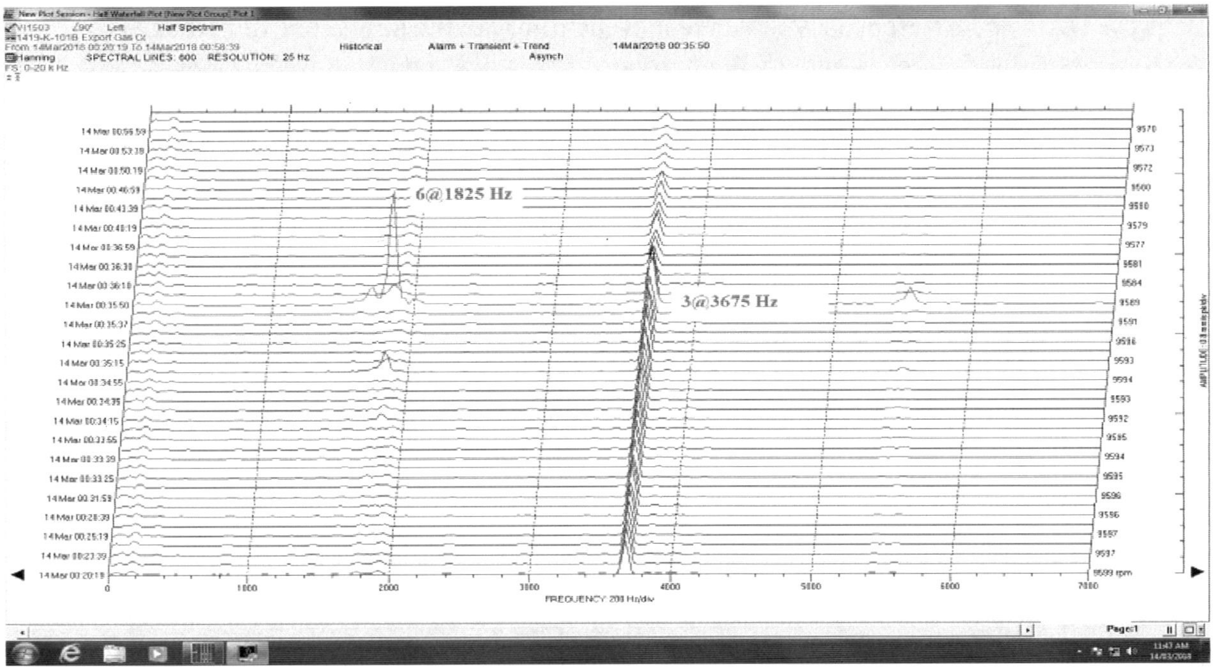

Fig 18 – Presence of high frequency component

It was checked if gear mesh frequency (GMF) or its multiples are exciting casing resonant frequency. It was found that constant peak of 3675 Hz was close to gear mesh frequency – 3700 Hz at that operating speed. Hence it was concluded that gear box structural natural frequency was getting excited at that operating frequency. For having further assurance, a heavy steel block was placed for quick test and it was found that casing vibration was subsided at that frequency. It was decided jointly that there is no value addition to change the stiffness of casing as it may lead to excitation at other operating frequencies.

High energy pump unit gear box vibration issue

A 12 MW direct online (DOL) driven water injection pump equipped with high speed double helical gear increaser used to show with high casing vibration and thus caused trouble to commission the pumps at site. It was checked whether we have an issue with grounding and filter settings for seismic sensors also were checked.

OEM responded as it is normal for a gearbox to show high vibration on startup when the power is turned on for heavy duty DOL start motor. There is a torque reaction between the gear teeth when power is applied and the shafts lift in their bearings, often to top clearance. This will give a large displacement in the shaft vibration and will show a high peak velocity in the seismic housing vibration. This usually lasts until a torque equilibrium is reached at a constant speed. It was noted in the plot that the vibration starts dropping as the unit starts to get up to speed, It was argued and thought on the fact that Bull gear and pinion DE and NDE shaft displacements found to be normal.

When FFT of casing vibration was reviewed, it was seen that we have the three main spikes on both machines are: close to 0Hz, 50Hz and 100Hz. Since the machine is only at 143rpm the first one is the DC/ very low frequency movement of the rotor i.e. the torque reaction, the next two are the supply, line frequency and 2X line frequency. The next bunch of frequencies is most likely gear meshing, whose frequency is rapidly changing with change in speed, and structural noise.

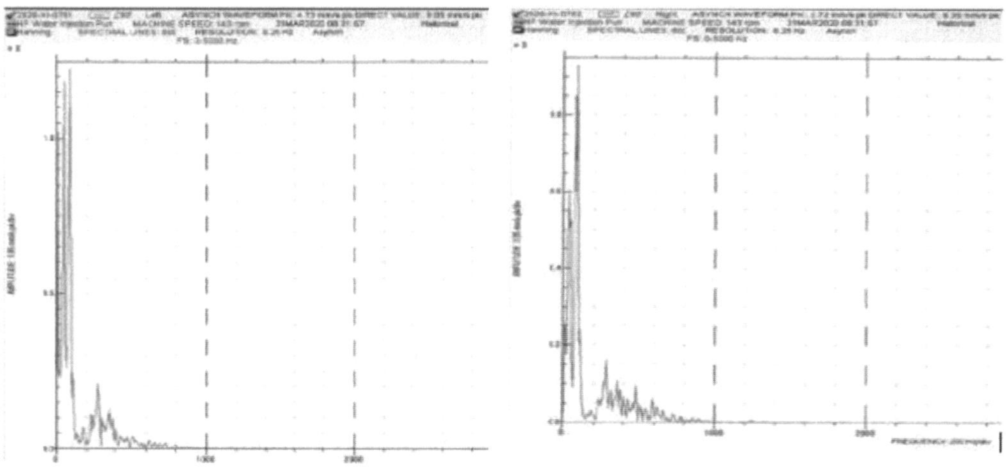

Fig 19: Spectrum of casing vibration

With comparison of gear box bull gear and pinion shaft displacement and no co relation was found with displacement and gear meshing frequency hence, it was concluded that such casing vibration was transient in nature and benign due to duration and severity. OEM incorporated a trip multiplier for 30 secs of start up in vibration monitoring panel.

Conclusion :

With deep delving into the matter , the issues should be investigated with detail RCA tools and then only should be concluded as benign and non-intrusive to machinery integrity.

Author ends up with making a point for requirement of deeper understanding to address such vibration phenomena during no load / coast down to decide whether issue to be considered as genuine or benign.

There lies a statement of caution to the end user / contractor / OEM should not rush with conclusion just to resume operation .This could always be a pragmatic aspect of recommendation such as tweak some software logic / include a trip multiplier than to deeply plunge into a design change proposal on such issues .

References –

1. A Review of Tilting Pad Bearing Theory - Timothy Dimond, Amir Younan, and Paul Allaire , International Journal of Rotating Machinery Volume 2011, Article ID 908469,
2. Nonsynchronous tilting pad bearing characteristics - Joachim Schmied et al, Delta JS, Proceedings of the 8th IFToMM International Conference on Rotor dynamics

3. HANDBOOK of LUBRICATION and TRIBOLOGY Application and Maintenance SECOND EDITION - Edited by George E. Totten Published in 2006 by CRC Press Taylor & Francis Group
4. Journal Bearing Vibration and SSV Hash - DeCamillo, Scan M.; He, Minhui; Cloud, C. Hunter 37th Turbomachinery symposium Texas A&M.
5. Rotor Instability Problems in An Integrally Geared Compressor Supported by Tilting Pad Bearings- Gruntfest, Peter M.; Andronis, Leo; William D Marscher, 30th Turbomachinery symposium Texas A&M
6. Torsional vibration of machines with gear errors - A W Lees, M I Friswell, G Litak Vibratory Power Flow Analysis of a Gear-Housing-Foundation Coupled System - Yafeng Ren, Shan Chang, Geng Liu, Liyan Wu, and Haiwei Wang. Shock and Vibration Volume 2018, Article ID 5974759, https://doi.org/10.1155/2018/5974759
7. Theoretical response on lateral response due to torsional excitation of geared rotors - J. S. Rao, T. N. Shiau and J. R. Chang Mechanism and Machine Theory Volume 33, Issue 6, August 1998,
8. Vibration and Noise Radiation Characteristics of Gear Transmission System- Jianxing Zhouand Sun Wenlei- Journal of low frequency noise vibration and active control. 33 No. 4 2014
9. Gearbox-monitoring-tutorial by Wilcoxon
10. Gearbox Spectral Components and Monitoring Methods- Presented by: Timothy S Irwin, P.E. Machinery Management Engineer Bently Nevada, Machinery Diagnostics Group
11. Vibration Analysis for Gearbox Casing Using Finite Element Analysis- Vijaykumar et al (IJES) ||Volume|| 3 ||Issue|| 2 ||Pages|| 18 - 36 || 2014 ||
12. Some Unusual Cases of Rotor Instability-Karl Olsson ASME Journal of vibration and acoustics Volume 125, Issue 4 , October 2003
13. An Independent Refiner's Approach - Practical Solutions to Complex Gear Problems. - John B Cary, 19th Turbomachinery symposium Texas A&M

Acknowledgements:

Author tenders his sincere gratitude to Mr. Ganesh Kumar Head of Mechanical department and Mr. Viljoen Christo, Head of integrity assurance Petrofac International for reviews and suggestions. Author is deeply thankful for support from MFPT Chairman and board of directors for review and accepting this paper for presenting in MFPT 2020 annual conference.

TWO-PLANE BALANCING METHOD FOR CORRECTION OF SHAFT UNBALANCE

Nazli Javadi Eshkalak[1], Hao Lu[1, 2], Chao Hu[1, 2], Steve Daining, and Dillon Wirth

[1] Department of Mechanical Engineering, Iowa State University, Ames, IA 50011

[2] Department of Electrical and Computer Engineering, Iowa State University, Ames, IA 50011

Abstract: Vibrations due to unbalance adversely affect the performance of rotating machinery. Hence, monitoring and minimizing such vibrations is important to prevent consequences, such as unexpected damage and catastrophic failures. This study presents numerical and experimental investigations of two-plane balancing to suppress the vibration of a rigid rotating shaft using an influence coefficient method. To examine the influence coefficient method, a test stand containing a rotating shaft supported by two ball bearings is provided. Balancing is performed by adding a set of weights onto two circular balancing planes attached to the bearings. Balancing experiments are conducted to evaluate the performance of the proposed method by utilizing the sensing and data acquisition hardware that consists of two accelerometers (one per bearing), a tachometer, and a data acquisition system. For experiments where multiple balancing iterations are required to achieve a balanced state, an influence coefficient matrix (ICM) refinement approach is developed. The refinement approach uses the data collected in the previous iterations to update the ICM so that the rotor can be balanced without putting more trial weights on either plane.

Keywords: Influence coefficient matrix; trial/correction weights; two-plane balancing; unbalance; vibration; weight splitting

1 Introduction

Vibrations resulting from mass/rotor unbalance are a major problem in rotating machinery. Thus, monitoring the positional changes and vibrations of rotating machine components (e.g., rotating shafts) is essential to ensure the operational reliability of bearings that often have shorter lifespans than other rotating machine components. Early fault detection and mitigation are critical in preventing catastrophic failures arising from excessive vibrations [1]. Balancing is considered an effective way to reduce vibrations caused by rotor unbalance and involves the process of adding/removing certain weights at specific locations on the rotor [2, 3]. To perform balancing, the maximum vibration threshold for each plane is defined, below which the system is considered balanced. Hardware such as tachometers, accelerometers, and data acquisition systems can be utilized to collect the vibration and rotor speed information for balancing [1]. It is necessary to choose robust and time-efficient balancing techniques. Field balancing [4, 5], also referred to as in-place balancing, is one technique that does not require removing any component from a machine for balancing purposes. An important advantage of field balancing is that an unbalanced rotating component can be balanced without removing it from the machine (i.e., with minimum disassembly and a low risk of affecting the assembly tolerances). Field balancing can be categorized into different procedures according to the type of designed rotor, viz. rigid or flexible.

It is noteworthy that the methods designed for rigid-rotor balancing may not be directly applicable to flexible rotors. Balancing methods can be broadly classified into two main groups: modal balancing methods for flexible rotors [6, 7, 8] and influence coefficient balancing methods for rigid and flexible rotors [9, 10]. Based on the number of balanced planes, the balancing methods can also be categorized into two main groups: single-plane balancing methods [11, 12] and multi-plane balancing vector methods [13, 14, 15]. In practice, multi-plane balancing is usually implemented as two-plane balancing, because a rotating element (e.g., a shaft) is typically supported by two bearings and has two balancing planes where operators can add/remove weights and accelerators that can measure vibrations.

Influence coefficient balancing methods are among the most widely used techniques for two-plane balancing [16]. According to the influence coefficient theory [17], vibration due to an unbalance is proportional to the unbalance. To represent the constant of proportionality between vibration and unbalance, a matrix containing complex numbers named influence coefficient matrix (ICM) is introduced, which can in turn capture the vibrational cross-effects and couplings between two balancing planes. The graphical solution method [18] is a basic approach for rotor balancing using the influence coefficient method. Generally, the graphical solution method plots the initial and trial vibration vectors, and then based on the graphical information it determines the correction weights. Carlson et al. [19] proposed a graphical method for single plane balancing, which did not require measurements for a phase response. Everett [20] expanded the single-plane balancing graphical method for a two-plane balancing problem, which could calculate the correction weights by making seven runs of the rotor. Mohamad et al. [21] proposed an ICM-based balancing method, and experimentally verified its performance for both single and two-plane balancing. It was observed that the vibrations of the rotor system were considerably reduced by applying the ICM-based method. Yu et al. [22] proposed a general ICM-based balancing method, which used data collected from a group of trial runs for each balancing iteration to refine the influence coefficients. Xu et al. [23] developed a dynamic balancing model by introducing a cross-correlation analysis method to extract the phase and amplitude information from the raw vibration signal; subsequently, they used the extracted information to calculate the ICM and correction weights.

We propose a two-plane balancing vector method using an ICM. The ICM is calculated based on the vibration responses upon adding different trial weights and is used to determine the correction weights required to offset the initial vibrations of a rotor system. However, when balancing a rotor system, the system may have some constraints that do not allow for the direct addition of the exact calculated correction weights, hence causing some residual vibrations. To address these limitations and minimize the residual vibrations, Huang et al. developed the weight splitting technique [10] by applying the least squares method to approximate an optimally calculated correction weight with two or more weights. The approximate correction weights obtained from weight splitting are then added to the balancing plane to minimize the residual vibration of the shaft. The present study employs the weight spitting technique and introduces a new approach to refining/updating an ICM for severely unbalanced cases where multiple balancing iterations are required. The effectiveness of the proposed two-plane balancing method was experimentally verified under a variety of unbalance settings. This paper is organized as follows. Section 2 provides background information on signal analysis and ICM based two-plane balancing. Section 3 describes the design of experiments and experimental results. The paper is concluded in Section 4.

2 Background

2.1 Signal Analysis

The vibration information of both planes would be required to monitor the balance condition of the rotor system. The vibrations due to rotor unbalance are highly dependent on the shaft speed. Therefore, the signal analysis algorithm is composed of a tachometer signal analysis algorithm and a vibration signal analysis algorithm.

The tachometer signal analysis algorithm is used to analyze the number of times the tachometer sensor detects the label, which indicates the number of rotations of the plane. First, the algorithm finds all time steps when the sensor detects the label by identifying when the voltage reading of the tachometer is less than a given threshold, labeled as $Threshold_1$ in Figure 1(b). In this study, the voltage threshold is set to 1 V. Note that one complete rotation has multiple time steps when the sensor detects the label. Hence, a time-interval threshold, labeled as $Threshold_2$ in Figure 1(b), is used to constrain the minimal time interval between the start of one rotation and the start of the next rotation. These two steps are performed on a 5-sec tachometer signal to estimate the time interval for each complete rotation in the signal. Finally, the speed of the shaft is calculated based on the average time interval per rotation.

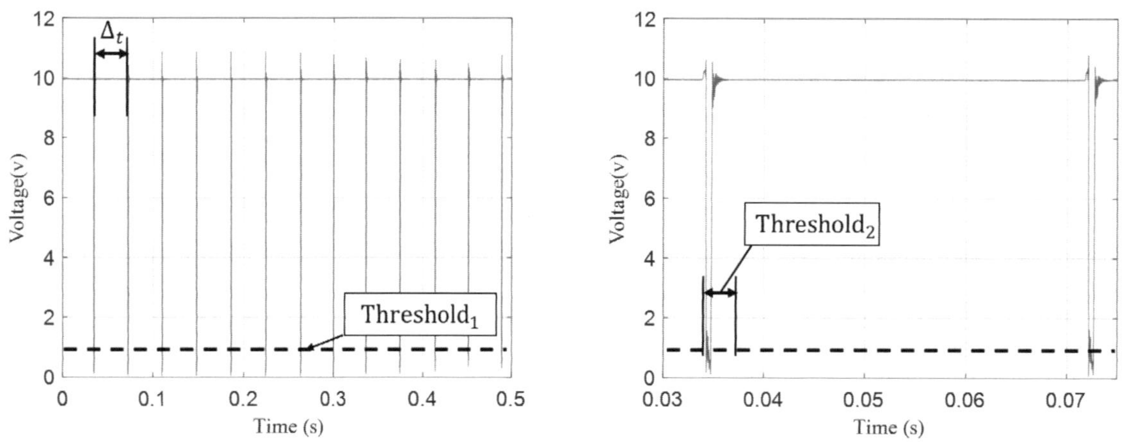

(a) (b)

Figure 1. Tachometer signal

Figure 2 shows the vibration analysis setting. The tachometer returns a low voltage signal whenc it senses the label. If a plane is unbalanced, its center of mass is not aligned with the center of rotation, which results in an unbalance vibration. Assuming there is a mass (located 90° counter-clockwise from the tachometer label) making the plane unbalanced, Figure 2(b) shows the relationship between the theoretical vibration and the phase for one rotation. Here, the phase is defined as the angle (in degrees) offset between the tachometer label and its initial position.

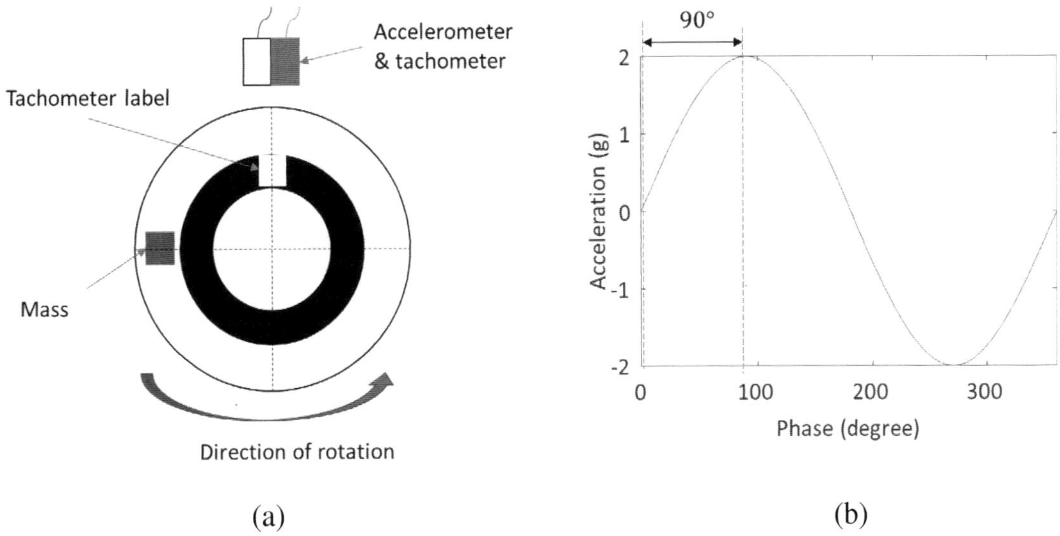

Figure 2. Arrangement for vibration measurement

Since a rotor system comprises multiple components and the collected acceleration signal is noisy, it is impracticable to extract the phase and amplitude of vibration directly from the raw signal. Note that the frequency of unbalance-induced vibration is equal to the shaft speed. Here, a bandpass filter that is implemented in the MATLAB built-in function "bandpass" is applied to denoise the vibration signal. The central frequency of the filter is set as the shaft speed, and the bandwidth is set to 2 Hz (i.e., 1 Hz on either side). An example of vibration signal processing is shown in Figure 3. After bandpass filtering, the raw signal (shown in Figure 3(a)) becomes a clean sinewave signal (in Figure 3(b)). For each rotation, the maximum acceleration value is considered the amplitude value. The time between the start of each rotation and when the maximum acceleration value is measured can be used to calculate the phase value.

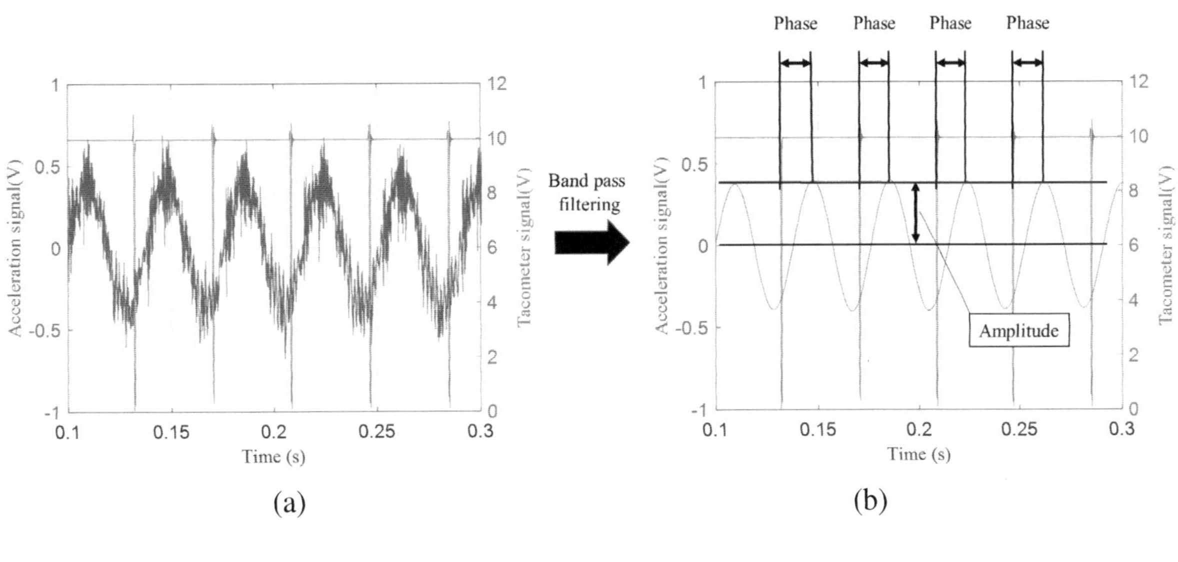

Figure 3. Vibration signal processing

2.2 Two-Plane Balancing

In comparison to single-plane balancing, two-plane balancing is more complex since the indications of unbalance at a particular plane do not represent the unbalance condition in that plane. This means that the simultaneous effects of unbalance forces on both planes should be considered when determining corrective actions. The aforementioned phenomenon, known as the cross-effect, makes two-plane balancing more complex than single-plane balancing. To quantify the cross-effect between two planes, the ICM would be required. Basically, the ICM represents how the unbalance on one plane affects the vibration on both planes.

2.2.1 ICM

Assuming that the system under consideration for balancing purposes is linear, the following equation represents the alterations of vibration signals as a result of weight placement [5]:

$$\begin{bmatrix} \vec{h}_{11} & \vec{h}_{12} \\ \vec{h}_{21} & \vec{h}_{22} \end{bmatrix} \begin{bmatrix} \overrightarrow{W_1} \\ \overrightarrow{W_2} \end{bmatrix} + \begin{bmatrix} \vec{A}_1^{(1)} \\ \vec{A}_2^{(1)} \end{bmatrix} = \begin{bmatrix} \vec{A}_1^{(n)} \\ \vec{A}_2^{(n)} \end{bmatrix} \quad (1)$$

where $\vec{A}_1$ and $\vec{A}_2$ are the vibration vectors on planes A and B, respectively, $\overrightarrow{W_1}$ is the weight attached to plane A, $\overrightarrow{W_2}$ is the representative of the weight being installed on plane B, and $\vec{h}_{11}$,$\vec{h}_{12}$,$\vec{h}_{21}$, and $\vec{h}_{22}$ form a 2×2 ICM. Superscript (1) denotes measurements without weights (i.e., the initial vibration of the system at the beginning of each iteration), whereas superscript (n) $(n \geq 2)$ denotes measurements with weights.

The two-plane balancing method requires at least three trial runs to obtain enough information for vibration analysis and suppression. The procedure of the three-step two-plane balancing method is briefly explained as follows:

1. Start the machine and collect the initial vibration information of planes A and B (using one accelerometer per plane and a tachometer).
2. Select a trial weight and attach it to the first plane (plane A). Record the weight and phase information of the trial weight for later calculations. Turn on the machine again and collect the trial vibration information of planes A and B.
3. Remove the trial weight and install it at the same phase angle on the other balancing plane (plane B). The machine will start again, and the amplitude and phase angle will be recorded.

After completing the aforementioned steps, one would be able to calculate the ICM and, consequently, the correction weights (for the first iteration of balancing). Based on Eq. (1), the 2×2 complex matrix of influence coefficients can be determined by:

$$\begin{bmatrix} \vec{h}_{11} & \vec{h}_{12} \\ \vec{h}_{21} & \vec{h}_{22} \end{bmatrix} = \begin{bmatrix} \vec{A}_1^{(2)} - \vec{A}_1^{(1)} & \vec{A}_1^{(3)} - \vec{A}_1^{(1)} \\ \vec{A}_2^{(2)} - \vec{A}_2^{(1)} & \vec{A}_2^{(3)} - \vec{A}_2^{(1)} \end{bmatrix} \begin{bmatrix} \overrightarrow{W_1}^{(2)} & \overrightarrow{W_1}^{(3)} \\ \overrightarrow{W_2}^{(2)} & \overrightarrow{W_2}^{(3)} \end{bmatrix}^{-1} \quad (2)$$

where $\vec{h}_{ij}$ ($i = 1, 2$ and $j = 1, 2$) is the element of the ICM representing the influence coefficient for the vibration of the ith plane due to a trial weight installed on the jth plane. $\vec{A}_1$ and $\vec{A}_2$ are the vibration vectors on planes A and B, respectively, $\overrightarrow{W_1}$ is the trial weight attached to plane A, and

$\overrightarrow{W_2}$ is the representative of the trial weight being installed on plane B. Furthermore, superscript (1) represents measurements without trial weights (i.e., initial conditions), and superscripts (2) and (3) respectively denote measurements with the first and second sets of trial weights. Once the ICM has been determined, the following formula can be utilized to calculate the correction weights required to offset the initial vibrations on both planes:

$$\begin{Bmatrix} \overrightarrow{W_1^{(c)}} \\ \overrightarrow{W_2^{(c)}} \end{Bmatrix} = \begin{bmatrix} \overrightarrow{h_{11}} & \overrightarrow{h_{12}} \\ \overrightarrow{h_{21}} & \overrightarrow{h_{22}} \end{bmatrix}^{-1} \begin{Bmatrix} -\vec{A}_1^{(1)} \\ -\vec{A}_2^{(1)} \end{Bmatrix} \tag{3}$$

Here, $\overrightarrow{W_1^{(c)}}$ and $\overrightarrow{W_2^{(c)}}$ are the calculated correction weights to be placed on planes A and B. The addition of these weights will generate vibration vectors that are similar in magnitude to $\vec{A}_1^{(1)}$ and $\vec{A}_2^{(1)}$, but in the opposite directions; this eliminates the initial vibrations of the rotor.

2.2.2 Weight Splitting

Owing to some practical constraints in rotating machine systems, achieving an optimal balancing performance becomes a challenge in terms of manufacturing procedures. For many balancing cases, there are limited numbers, positions (areas to place weights on the balancing planes), and mass of balancing/correction weights. Consequently, it is not feasible to attach the exact calculated correction weights onto the planes for the balancing purpose. This will, in turn, cause some residual vibrations in the rotor system even after completing the balancing process. These restrictions make it necessary to apply an approximation to the calculated correction weight for each plane. This approximation involves "weight splitting" for the available hole-weight combinations. In this case, all combinations of candidate holes and weights would be utilized for the weight splitting.

In summary, the proposed weight splitting approach was based upon finding the pairwise combinations of all/specified weights and holes on the balancing planes. In other words, all available candidate hole-weight combinations should be considered for weight splitting. However, in the presence of a further restriction being introduced to reduce the computation time for each balancing iteration, a maximum of two weights per plane are considered for approximating correction weights. The resulting correction weight vector is obtained by combining two vectors of split weights (i.e., the vector sum). Hence, these two assumptions/conditions introduce more restrictions in terms of calculations.

Once all possible hole-weight combinations for weight splitting are determined, the least squares algorithm can be implemented to find the combinations of maximum two split weights, where their resulting vector sum is closest to the calculated correction weight. To apply the least squares algorithm, the following l_2 vector norm was selected as the objective function to be minimized [10]:

$$\min_{x_{jk}, j=1,2,\ldots,n_h,\ k=1,2,\ldots,n_w} \left\| \begin{matrix} \sum_{j=1}^{n_h} \sum_{k=1}^{n_w} x_{jk} w_k \cos(\beta_j) - \mathrm{Re}\left(W_i^{(c)}\right) \\ \sum_{j=1}^{n_h} \sum_{k=1}^{n_w} x_{jk} w_k \sin(\beta_j) - \mathrm{Im}\left(W_i^{(c)}\right) \end{matrix} \right\|_2 \tag{4}$$

where x_{jk} is the mass indicator (being 0 when there is no weight on a hole and 1 when the hole is occupied by a set screw), n_h and n_w are the number of holes and weight sizes, respectively, w_k is the amount of weight for the kth weight type, β_j is the phase of the jth candidate hole being occupied by split weight, and $W_i^{(c)}$ is the calculated correction weight on the ith plane ($i = 1$ plane A and $i = 2$ plane B). Figure 4 illustrates the least square method for weight splitting purposes, where the red arrow and red circle mark the direction of the calculated correction mass vector and the location of the correction mass, respectively, while the black solid circles mark the optimized locations of the split weights.

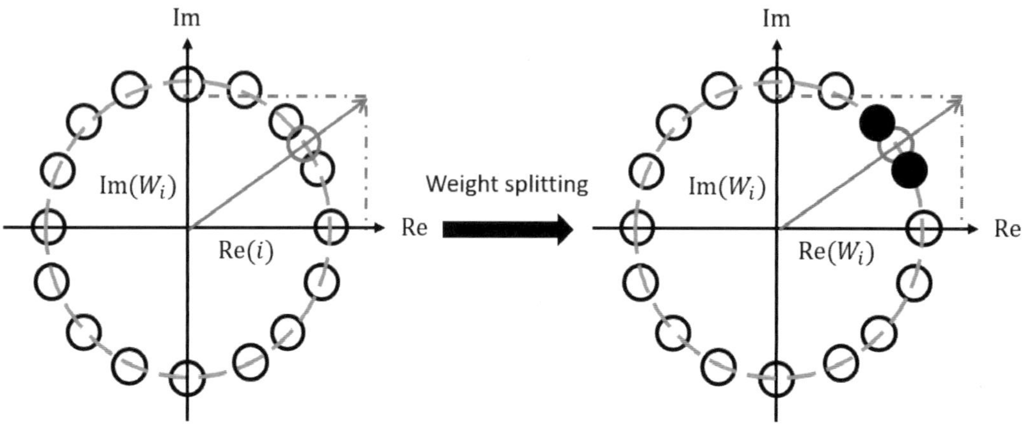

Figure 4. Weight splitting by optimizing/minimizing l_2 vector norm.

2.2.3 ICM Refinement

For some severely unbalanced situations, several iterations might be required to balance the rotor. Further, for each balancing iteration, the ICM should be updated to identify the correction weights. In the present study two approaches were considered for updating the ICM:

- **Approach 1:** Recalculate an ICM in a way similar to the first iteration (Eq. (2)). Perform the three-step balancing procedure by adding two trial weights to the planes and calculate a new ICM based on the resultant vibrations. In this approach, the initial vibration of the nth iteration would be residual vibration of the $(n-1)^{th}$ iteration.
- **Approach 2:** Refine an existing ICM at each iteration (e.g., at the nth iteration) based on the most recent vibration data and correction weights (e.g., at the $(n-1)^{th}$ iteration), without a second use of trial weights, using the generalized least squares method. Essentially, it is unnecessary to repeat the trial-weight addition procedure at each iteration (except for the first iteration to define the ICM). The updated ICM will be saved for updating in the next iteration (e.g., the $(n+1)^{th}$ iteration).

The benefit of the first approach is that it accounts for any changes in rotor dynamics, while that of the second approach is that it increases the sample size and presumably increases the accuracy in the absence of precision.

As illustrated in Figure 5, for refining the ICM (Approach 2), one should do the three-step balancing only for the first iteration and save the resultant ICM for later iterations. In other words, after obtaining the ICM, as well as calculating and placing the correction weights at the first iteration, these correction weights can be considered as a new set of trial weights with new resultant vibrations and hence can be utilized to update/refine the existing ICM. This procedure is repeated to balance the rotor system at each iteration. It can be observed in Figure 5 that, similar to Approach 1, the residual vibrations upon adding the correction weights to the system at the $(n-1)^{th}$ iteration are considered as the initial vibrations of the nth iteration; however, in contrast to Approach 2, it is unnecessary to attach trial weights in each iteration (except for the first iteration, when the ICM is not already defined).

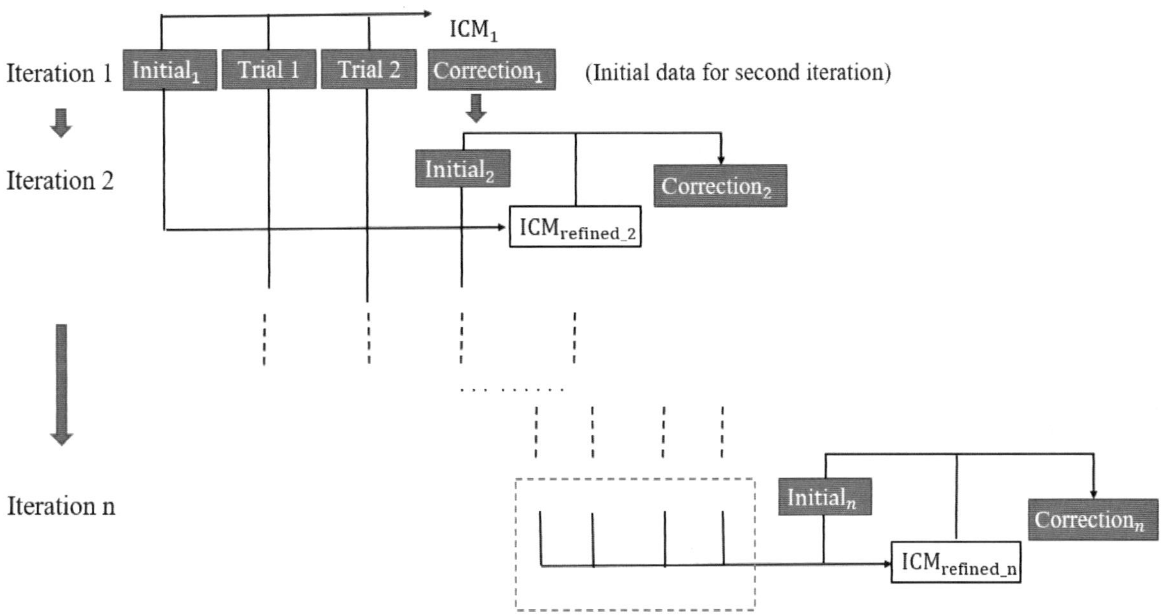

Figure 5. Refining the existing ICM based on the most recent data without a second use of trial weights

3 Experimental Validation

A set of experiments were conducted on a test fixture to evaluate the performance of the proposed ICM-based two-plane balancing method. Figure 6 shows the balancing test stand for this study. Seventy-two equally spaced holes around the circumference of the two balancing planes and six different balancing weight sizes (3.396, 2.135, 0.867, 0.491, 0.231, and 0.095 grams) were used for the weight splitting and balancing purposes. The speed of the shaft was set at 26 Hz. Each candidate hole on the test stand can only hold one weight, which means that if a hole is occupied by a weight (set screw), it will no longer be available for subsequent weight splitting processes.

Figure 6. Test fixture for conducting the two-plane balancing experiments

Six single-plane unbalance settings and six dual-plan unbalance settings were introduced to evaluate the performance of the proposed two-plane balancing method. Table 1 summarizes the unbalance settings of the designed experiments for this study.

Table 1. Summary of unbalance settings used in the experiments

Unbalance setting		The unbalance weight and location
Single-plane unbalance (on plane A)	1	3.396 g @ 125°
	2	2.136 g @ 245°
	3	2.136 g @ 125° and 3.396 g @ 130° and 2.136 g @ 135°
	4	3.396 g @ 125° and 3.396 g @ 130° and 2.136 g @ 245° and 2.136 g @ 250°
	5	3.396 g @ 125° and 3.396g@ 135° and 3.396 g @ 155°
	6	3.396g @ 125° and 2.136 g @ 130°and 3.396 g @ 135°
Dual-plane unbalance (on planes A and B)	7	3.396 g @ 125°
	8	2.136 g @ 245°
	9	2.136 g @ 125° and 3.396 g @ 130° and 2.136 g @ 135°
	10	3.396 g @ 125° and 3.396 g @ 130° and 2.136 g @ 245° and 2.136 g @ 250°
	11	3.396 g @ 125° and 3.396g@ 135°and 3.396 g @ 155°
	12	3.396g @ 125° and 2.136 g @ 130°and 3.396 g @ 135°

To conduct the two-plane balancing method, in addition to the weight and location settings utilized to introduce an initial unbalance condition on the system (Table 1), a unique set of trial weight and location was selected and used throughout all the experiments to calculate and update the ICM and hence calculate the correction weights (regardless of the approach for updating the ICM). Regarding Approach 1, the selected trial-weight and location setting was used for each iteration to recalculate the ICM; whereas for Approach 2, the same selected trial-weight and location setting was only utilized in the first iteration to calculate the ICM.

The vibration vectors of planes A and B, before and after employing the two-plane balancing method using Approach 1 for updating ICM, were collected and are shown in Table 2. As shown in Table 2, after installing the correction weights, the vibration amplitude of the unbalanced plane decreases significantly at each balancing iteration. For some severe unbalance settings (e.g., settings 3 and 9), multiple iterations were required to reduce the mostly unbalance-induced vibrations to acceptable levels. For all the unbalance settings, the test stand could be balanced within two balancing iterations.

Table 2. Balancing performance using Approach 1

Unbalance setting	Number of iterations	Before putting correction weights				After putting correction weights			
		Plane A		Plane B		Plane A		Plane B	
		Amplitude (g)	Phase (°)	Amplitude (g)	Phase (°)	Amplitude (g)	Phase (°)	Amplitude (g)	Phase (°)
1	1	0.1589	135.1	0.0135	242.3	0.0101	152.2	0.003	160.5
2	1	0.0791	248	0.0178	50	0.0025	229.7	0.0128	85.7
3	1	0.2942	133.2	0.0382	249	0.032	149.1	0.0053	196.6
3	2	0.032	149.1	0.0053	196.6	0.0025	149.9	0.0001	140.9
4	1	0.2321	168.6	0.0311	298.5	0.0096	172.8	0.0051	249.8
5	1	0.3669	142.5	0.0615	266.9	0.1091	0	0.0219	257.1
5	2	0.1091	141.6	0.0219	257.1	0.0078	145.7	0.0015	233
6	1	0.3371	133.9	0.0526	254.1	0.077	139.7	0.0132	239.4
6	2	0.077	139.7	0.0132	239.4	0.008	78.8	0.0053	189
7	1	0.1244	135.4	0.1403	130.9	0.0068	100.6	0.0088	83.5
8	1	0.0775	256.3	0.0723	250.6	0.0059	256.2	0.006	266.7
9	1	0.2715	143.5	0.2963	139.4	0.0338	119.1	0.0498	112.3
9	2	0.0338	119.1	0.0498	112.3	0.001	282.9	0.0013	168.9
10	1	0.2157	179.5	0.2298	174.2	0.0264	280.7	0.0139	282.3
10	2	0.0264	280.7	0.0139	282.4	0.0037	292.6	0.0107	199
11	1	0.3296	153.6	0.3576	149.3	0.1088	164.4	0.1227	156
11	2	0.1088	164.4	0.1227	156.1	0.0052	34.2	0.0034	222.6
12	1	0.3066	144.7	0.3327	140.7	0.0753	148.8	0.0885	140.2
12	2	0.0753	148.8	0.0885	140.3	0.0094	236.4	0.0166	221

To evaluate the performance of the two different approaches for updating ICM, a new set of experiments were designed and executed. In these experiments, unbalance settings 5, 6, 11, and 12 were used to generate several severe unbalance situations. For the second balancing iteration, the two ICM updating approaches (Approaches 1 and 2) were applied. The changes of vibration vectors are shown in Table 3. As demonstrated in Table 3, both approaches can reduce the residual vibrations significantly. After the second balancing iteration, the residual vibrations on planes A and B became comparable.

Table 3. Comparison of balancing performance between two different ICM updating approaches

Approach for updating ICM	Unbalance setting	Before putting correction weights				After putting correction weights			
		Plane A		Plane B		Plane A		Plane B	
		Amplitude (g)	Phase (°)	Amplitude (g)	Phase (°)	Amplitude (g)	Phase (°)	Amplitude (g)	Phase (°)
Approach 1	5	0.1107	153.7	0.0278	252.2	0.0049	193.0	0.0027	221.5
	6	0.0728	139.7	0.0177	227.1	0.0016	272.8	0.0006	210.4
	11	0.1170	171.4	0.1278	163.8	0.0068	202.3	0.0073	287.7
	12	0.0827	162.8	0.0918	153.9	0.0049	181.9	0.0045	279.4
Approach 2	5	0.0324	148.2	0.0128	215.7	0.0029	340.4	0.0005	22.0
	6	0.1107	153.7	0.0278	252.2	0.0019	260.4	0.0054	293.3
	11	0.0728	139.7	0.0177	227.1	0.0045	339.7	0.0016	294.0
	12	0.0477	184.1	0.0538	167.1	0.0134	352.3	0.0050	339.9

4 Conclusion

In this paper, a two-plane balancing method based on the use of influence coefficient matrix (ICM) was experimentally investigated. For the balance settings where multiple balancing procedures were required, a new ICM refinement approach was proposed to determine the correction weights without adding more trial weights. The proposed method is robust and easy to implement. Furthermore, the experimental results demonstrate that this method is capable of considerably reducing unbalance-induced vibrations on both balancing planes.

References

1. Wang, W., & Jianu, O. A. (2009). A smart sensing unit for vibration measurement and monitoring. IEEE/ASME Transactions on mechatronics, 15(1), 70-78.
2. Kalmegh, A., & Bhaskar, S. (2012). Dynamic balancing of centrifugal pump impeller. International Journal of Emerging Technology and Advanced Engineering, 2(6), 409-413.
3. Rodrigues, D. J., Champneys, A. R., Friswell, M. I., & Wilson, R. E. (2011). Two-plane automatic balancing: A symmetry breaking analysis. International Journal of Non-Linear Mechanics, 46(9), 1139-1154.
4. Lund, J. W., & Tonnesen, J. (1972). Analysis and experiments on multi-plane balancing of a flexible rotor. Journal of Engineering for Industry, 94(1), 233-242.
5. Yu, J. J. (2009). Relationship of influence coefficients between static-couple and multi-plane methods on two-plane balancing. Journal of engineering for gas turbines and power, 131(1).
6. Bingang, X., & Liangsheng, Q. (2001). A new practical modal method for rotor balancing. Proceedings of the Institution of Mechanical Engineers, Part C: Journal of Mechanical Engineering Science, 215(2), 179-189.
7. Saito, S., & Azuma, T. (1983). Balancing of flexible rotors by the complex modal method.
8. Deepthikumar, M. B., Sekhar, A. S., & Srikanthan, M. R. (2013). Modal balancing of flexible rotors with bow and distributed unbalance. Journal of Sound and Vibration, 332(24), 6216-6233.

9. Darlow, M. S. (2012). Balancing of high-speed machinery. Springer Science & Business Media.
10. Huang, B., Fujimura, D., Allaire, P., Lin, Z., & Li, G. (2009). Constrained balancing of two industrial rotor systems: Least squares and min-max approaches. Shock and Vibration, 16(1), 1-12.
11. Rodrigues, D. J., Champneys, A. R., Friswell, M. I., & Wilson, R. E. (2011). Experimental investigation of a single-plane automatic balancing mechanism for a rigid rotor. Journal of Sound and Vibration, 330(3), 385-403.
12. Sperling, L., Ryzhik, B., & Duckstein, H. (2004). Single-plane auto-balancing of rigid rotors. Technische Mechanik. Scientific Journal for Fundamentals and Applications of Engineering Mechanics, 24(1), 1-24.
13. Ozoegwu, C. C., Nwangwu, C. C., Uzoh, C. F., & Ogunoh, A. V. (2012). zPure analytical approach to rotational balancing. J Saf Eng, 1(4), 50-56.
14. Zhou, S., & Shi, J. (2001). Active balancing and vibration control of rotating machinery: a survey. Shock and Vibration Digest, 33(5), 361-371.
15. Dyer, S. W., Shi, J., Ni, J., & Shin, K. K. (2001). Robust optimal influence-coefficient control of multiple-plane active rotor balancing systems. J. Dyn. Sys., Meas., Control, 124(1), 41-46.
16. Ibraheem, A. A., Ghazaly, N. M., & Abd el-Jaber, G. T. (2019). Review of Rotor Balancing Techniques. American Journal of Industrial Engineering, 6(1), 19-25.
17. Kroon, R. P. (1944). Balancing of rotating apparatus, part ii. J. Applied Mechanics, 11, A47.
18. Kelm, R., Kelm, W., & Pavelek, D. (2016). Rotor Balancing Tutorial. In Proceedings of the 45th Turbomachinery Symposium. Turbomachinery Laboratories, Texas A&M Engineering Experiment Station.
19. Carlson, P.O.L., 1979, May. Four run balancing without phase. In Proceedings of Machinery Vibration Monitoring and Analysis Seminar and Meeting, New Orleans, LA (pp. 411-418).
20. Everett, L. J. (1987). Two-plane balancing of a rotor system without phase response measurements.
21. Mohamad, W. S. W., Mat Isa, A. A., & Ismail, M. A. (2013). Analytical and Experimental Study on Balancing Using Influence Coefficient Method. In Applied Mechanics and Materials (Vol. 393, pp. 694-702). Trans Tech Publications Ltd.
22. Yu, X., 2004. General influence coefficient algorithm in balancing of rotating machinery. International Journal of Rotating Machinery, 10.
23. Xu, X., & Fan, P. P. (2013). Rigid rotor dynamic balancing by two-plane correction with the influence coefficient method. In Applied Mechanics and Materials (Vol. 365, pp. 211-215). Trans Tech Publications Ltd.

MANAGEMENT OF STRESS – A MECHANICAL SYSTEM SIMULATION APPROACH

DR. J. PATTABIRAMAN,

Formerly Dean Mechanical engineering KCG college of Technology, Madras. ;
Manager (R &D) Tata consulting Engineers Bangalore;
Sr. Manager (CAE) Ashok Leyland Chennai , India
and Director Christ College of engineering and technology, Pondicherry , India

Introduction:

This paper is the result of several years of the author's work related to stress analysis, modeling and simulation for vibration and failure analysis. With experience gained in management and mechanical design & analysis, it is felt that there appears to be some similarity between human behavior and mechanical behavior of materials. The purpose of this article is to bring out the striking comparisons and commonalities and therefore suggest some practical methods to understand and manage human stress which is an essential requirement for assessing performance of today's professionals.

Basics of Mechanical Behaviour:

Loading of a specimen invariably causes mechanical stress. The material of the specimen is subjected to some strain and stress. Fibres or locations within the specimen moving or dislocating with respect to each other cause strain. If the body or specimen is free, the applied load causes only a gross displacement and the body moves as governed by Newton's laws of motion. If the body is constrained, or prevented from any movement, the points inside the body move relative to each other and cause a strain. ***It is the differential displacement within the body that is responsible for causing a strain.***

By virtue of material property, strains induce stresses. The body, which is constrained, resists the external load through development of internal stresses to keep itself in equilibrium. The same body can be treated as a rigid or an elastic body according to the situation. In case, the body is subjected to temperature rise, there is an expansion. If the expansion is unconstrained, then also there is no stress, although apparently there is a strain due to change in length.

Further, the external load acting on the constrained body causes stress and as the load increases, the stress also increases. The inherent property of the material permits the stress to increase proportionately with load upto some limit. Beyond that, either the material fractures (brittle material) or yields (ductile material). Stresses increase for some more time and then the material behaviour shifts to plasticity and undergoes large deformation and breaks. Most materials, after removal of load, exhibit permanent deformation.

All materials however contain inherent defects of manufacture, albeit that they could be extremely minute defects, that are not detectible by standard inspection processes. Favorable loading conditions and environments may exist to permit these defects to be contained. There could also be unfavorable loading conditions and adverse environments that could accelerate the growth of these defects. The defect grows and reaches a stage where the material is no longer able to carry the external load further and as a result, fractures. This is the basic principle underlying fracture mechanics, crack initiation and crack growth theories. In the case of alternating or cyclic loading, the material is subjected to fatigue and depending upon its endurance strength, it either survives or fails after some number of cycles.

The attention of the reader is drawn to specific words used so far like **Stress, Strain, Yielding, Rigidity, Elasticity or Flexibility, Defects, Defect Growth, Fatigue, Endurance & Failure** in the context of mechanical behaviour. These "Operative Words" also commonly used in describing human behavior and the human response to stress which is good parallelism to material behavior.

Similar terminology is also commonly used in electrical engineering with terms like Resistance, Admittance, Reluctance, Conductance, Impedance and Mobility.

Dynamic response:

The basic mechanical behavior discussed so far does not involve "Time" as parameter. In order to understand the human stress, there is a definite need to extend the system behaviour to dynamic loads or excitation. A simple vibration model consisting of spring-mass-damper elements often deals this with.

The spring is a storehouse of potential or strain energy. The mass represents inertial property that resists change (position or uniform motion). The damper provides control of movement. Excessive dynamic movement is brought down under control by the damping and helps the system to come to a position of Rest.

The attention of the reader is again drawn to words like **Energy, Inertia and Control**.

Parallelism with human response:

The key words identified in the foregoing discussions lend some **striking similarities between human behavior and mechanical behaviour of materials.**

The Human Being is a much more complex entity than Materials. Even in the case of materials, there is unpredictability in their behavior, but the degree of variation is only statistical. But in the case of human beings, the behavioural dynamics depends upon so many factors, the mind playing the foremost role (The word mind and brain are sometimes used interchangeably by behavioral scientists).

The stress here refers only to mental stresses, the body playing a responsive role through the mind. The pains caused to the body to a large extent, can be cured by medical sciences. The power of the Mind in terms of 'Will Power' to get cured and a 'Positive Mental Attitude' to achieve restoration of good health, play a crucial role to aid the medicines and pills to act in a favourable manner to bring down / cure the illness.

Disturbances and deviations from equilibrium status continue to occur in almost in all walks of human life. We witness people either over-reacting or being calm to being indifferent; while some prolong their anxieties and worries, a very few endure things with a smile and carry on with their work. ***What is the secret of success of these very few?***

There is a very thin line demarcating between the desirable and undesirable characteristics in behavior. Lifelong experiments go on continuously within the human mind, as to where to draw this line. Invisible thin lines of demarcation exist between words like *intelligent and clever, worry and concern, involvement and interference, adamant and determined, patience and delay, argument and discussion, effectiveness and efficiency.* The management experts will quite readily perceive the differences through the behavioral responses of a person. The differences are very subtle and perceived by colleagues, subordinates or superiors. It may even be difficult to spell out exactly or articulate well on the type of response.

Models-Mechanical vs Behavioral: A behavioral model, which resembles mechanical systems, is presented below with a parallel drawn between the human mind and mechanical system. The feedback loop represents actions for improvement. It may be emphasized that any improvement is a continuous process and gets reflected in behavioural changes for the better.

MECHANICAL SYSTEM RESPONSE MODEL

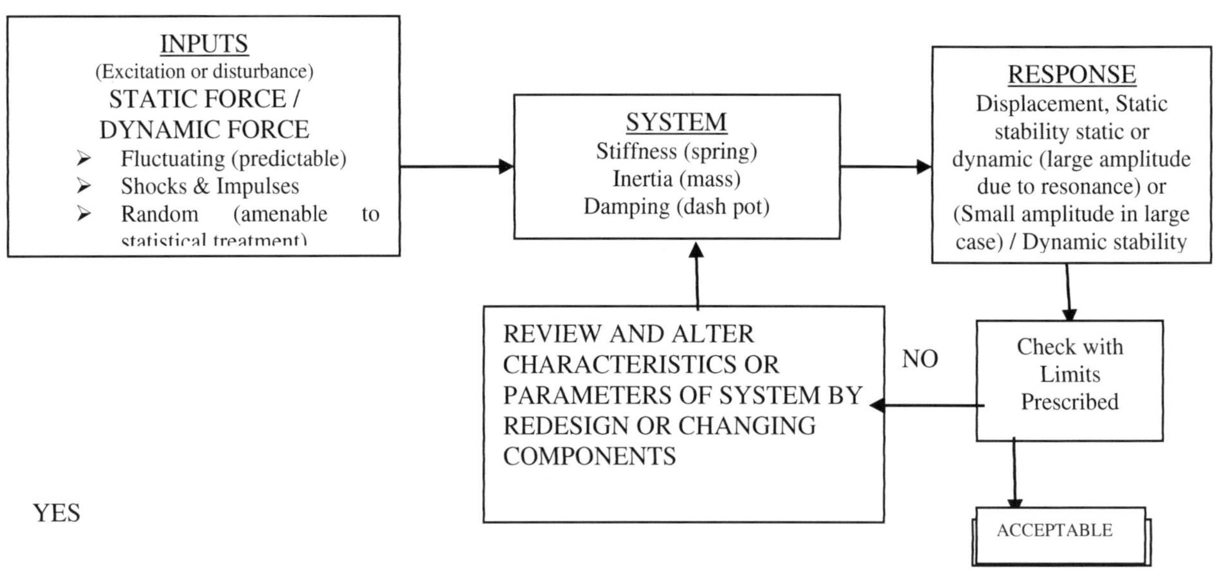

HUMAN SYSTEM RESPONSE MODEL

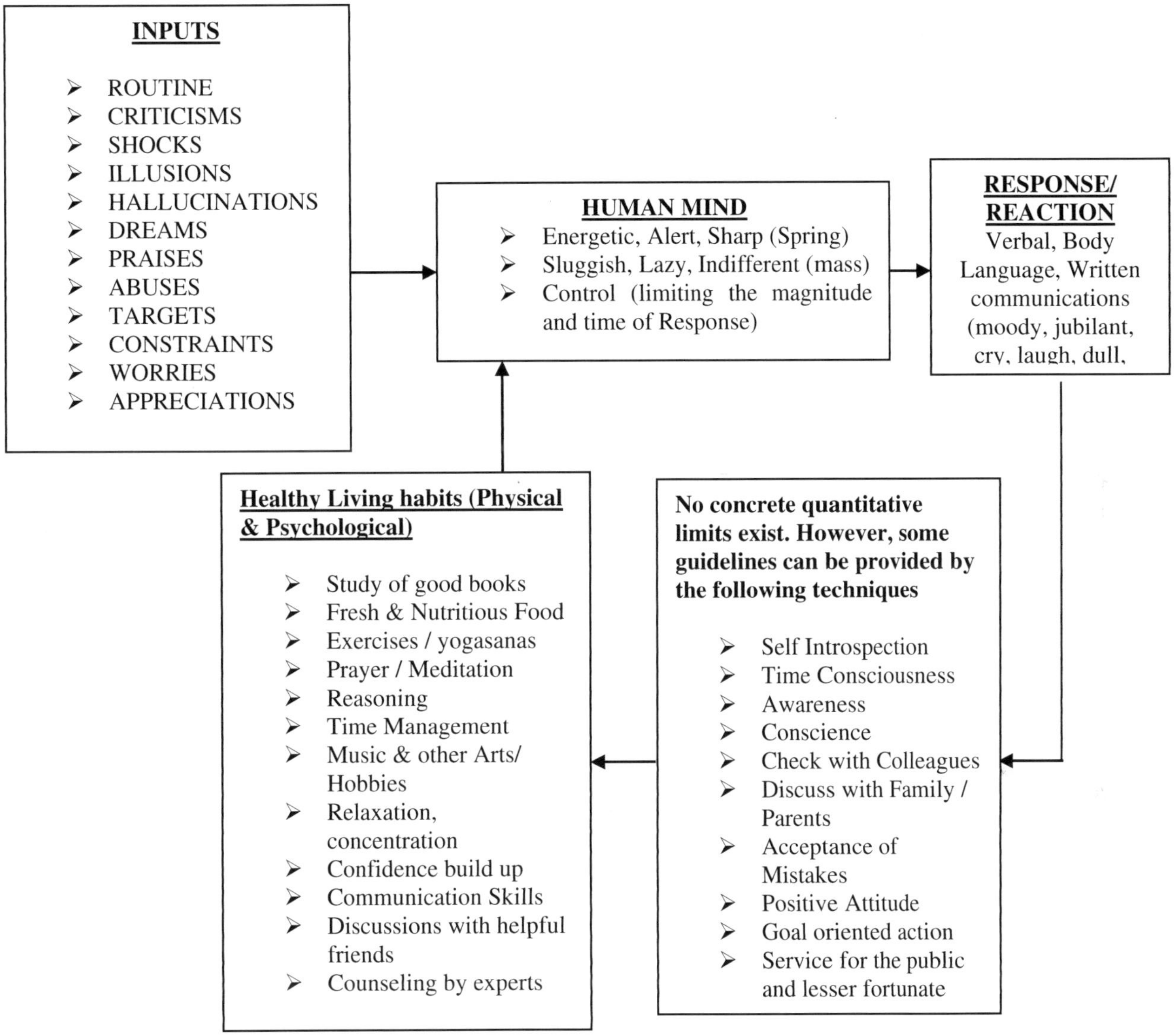

The essential difference in characteristics of inputs is that the human mind receives dynamic inputs all the time. The inputs are changing instant to instant and can be a function of space, location and the source of input. Depending on its strength, the mind processes these inputs, screens, assimilates, converts or rejects and produces a resultant response as illustrated in the block diagram.

It may please be noted that what comes to the human system, as input is a ***mere description of situation.*** The mind converts the situation into a problem or an issue. ***A problem is created by the mind when there is a difference between what the situation is***

and what is expected/desired by the mind. The understanding of the situation and assimilating it, is an independent process. Often, a confusion arises when the situation is not perceived in its entirety but coloured by the desired event. It is this difference in dynamics of the thought process, which is responsible for the mental strain leading to human stresses. Hence, if the situation is understood and accepted first calmly, the mind becomes freer and that itself will pave way for facing the situation and finding the right steps for solution.

If the situation faced is as per what was expected / desired by the mind, then the problem does not arise, as the differential thought movement in the mind between expectations and reality of the situation, is minimal. If this subtle principle is thought of, during the problem initiation stage itself, the problem can be arrested right at the beginning, without allowing it to grow.

Dynamics of the mind:

It often happens that work pressure or load (assigned by organisation or family or self) tends to overload the human system. It must be emphasized here that the capacity of the human brain is unlimited. Even world-renowned geniuses like Albert Einstein used only 4% of brainpower as against normal human beings who use only 2%. Hence there is nothing like overloading. It is purely a psychological phenomenon creating an illusion that we alone are doing that work or taking up that entire load. It is a load on the system. A careful study of the organisation structure reveals provision and place for departments, service centres, computer facilities, superior officers and management personnel who are all involved in sharing some part of the load with varying degrees of complexities. A proper planning and understanding of the total task will help in the appropriate breakup of tasks and proper load distribution. In case a portion of work lies beyond the present capability of the incumbent, it may be politely indicated at the earliest stage to the superiors, so that they could suggest alternatives like additional manpower, transfer of assignment or training to the existing persons or off-loading to an outside agency to ease the situation, but still working towards achieving the target. There would be no problem if this effort were taken at the earliest or in the initial stages itself. There will not be any surprises springing up towards a later stage or at the end.

As an analogy, in mechanical design, all members take up and share the load as per the load flow through the members, which are designed to carry the distributed load.

To understand the nature of mind, which is dynamic all the time, the problem arising from the mind can be explained with the following example.

Quite often, people say that they have done a work successfully expecting a reward or appreciation. It could as well happen just the opposite way. Take the example of a person proceeding out of station and returning to the same place with successful results and feels that 'he has' done the job. A careful analysis of the steps involved in this success story indicates the continuity of the following chain of events, which should have been successfully accomplished, if success in his mission is to be achieved.

1. He should get ticket and reservation
2. Train/ Flight should reach safely and in time
3. Local conveyance employed should be available to take him to his destination
4. Weather conditions should be favourable and permit his movement
5. Health should be favourable
6. Person in the place should be available, free and extend his cooperation
7. Support staff should cooperate to get the deliverables in time
8. Communication should be cordial and encouraging one
9. Return journey availability, safety and comfort.

Just as the modern management speaks of "supply chain management", the above list could be termed as "*Event Chain Management*".

Widening the Mind:

If this were the case, malfunction of any one event can jeopardize "success" of the entire mission. One must be aware that there is an unseen coordination between all activities to accomplish or assist in the mission and that should be if not openly, at least inwardly acknowledged. This implies that there are many events in the chain for which we have no control, and these are situations, which come together, by the way. Therefore, it must be understood how little our role is in the total process of achieving success. Such a feeling also will slowly reduce the illusory feeling that 'I have' accomplished. A steady shift in mental approach from "I & Mine" to "We and Ours" is a good approach to reduce the mental load and stress. In case the Event Chain is breaking at a point, one also knows and understands that it was due to an aspect/s, that were totally beyond his/her control and hence the mission was a failure with respect to the desired accomplishment of the result. While lots of reason and excuses can be given for failure, equally so, when it comes to success achieved in a mission, one should acknowledge and express the contribution of the various entities and forces that worked together inconspicuously. This is a typical example where a deeper analysis or introspection can lead to a change in system characteristics to result in a composed, considered and a mature behavioural response. The control element (damping) is tuned better to get the desired response, through self-analysis, self-introspection and contemplation.

Undesirable response – a behavioural accident:

For any accident to happen, there must be synchronization between time and space. If there is a variation in one of the space coordinates or time coordinate, the accident does not take place. Getting angry is a mental phenomenon and anger causes physical and mental discomfort and can estrange relations between persons or organizations.

Anger expresses itself at a time, location with persons/situations. When we watch our mind, we recognize when it gets excited. The best way to avoid the 'accident' is to let the mind become calm by postponing any immediate or impulsive response. When you allow a little time consciously, the same undesirable/unpleasant situation can be managed and conveyed more peacefully and effectively, rather than letting emotions overpowering the mind.

The result of mechanical impact on a structure can cause damage as the load bearing capacities may far exceed its mechanical strength. The result of mental impact has far reaching consequences. The matter on which anger was expressed is most often forgotten but only the angry tone and angry words and angry reactions are often remembered.

However, with a stable mind, applying proper controls, the mental state soon returns to balanced condition and emotion tends to get replaced by logic, understanding and acceptance of the situation. Taking due care of this aspect, conflicts likely to arise due to improper communication can be avoided at the earliest and stress on account of this, is largely minimized. The human mind unlike conventional structural materials, can be termed as "super smart structure" which can show a much-delayed response to an input.

Is desire unhealthy?

Ideals of life presented in scriptures and holy books like "desirelessness", "surrendering the fruits of action and not concerned with results" and "taking things as they come" are nice ideals but difficult to straightaway adopt in the present fast-moving, material conscious and time conscious world. Once the control of mind is practiced, the lofty ideals will become more adaptable and become a part of our actions, refine our system characteristics to reduce stress and enjoy harmony and peace. The mechanical system simulation approach gives a scientific and technical interpretation to understand to some extent, the rather complex behavioral model of the human mind. Moderate desires and those, which can be achieved with in the means and resources through honest and righteous methods, are healthy and provide a purpose to life. If the desires are in excess, it can lead to mental over burdens and can be a source for any disorder or incoherent behaviour.

Conclusion:

To professionals, it is extremely important to have a clear, healthy and stable mind to perform and excel in their duties and responsibilities. Watching the mind and being conscious of its movement itself is the first step in training towards controlling the mind. With control of the mind, concentration gets enhanced and the entire system, the body, the mind and the intellect coordinate so well and the task gets completed with a feeling of joy. If things have gone wrong, the first step is to fix the problem and set it right, then to go about investigating on what went wrong.

The management of stress, although seems to be a corporate issue in many human resource development departments, since the root of the problem is the individual's attitude to accept the situation first and thereafter change towards improving attitude and behavior. This is more applicable at leadership levels. When a leader enjoys inward happiness and peace, he radiates this to all around him and spreads and creates a very healthy, joyous and congenial environment to work for the organisation where the stresses and strains get minimized.

ADDITIVE MANUFACTURING WITH STEEL POWDER WHICH WAS GAS ATOMIZED IN A MOBILE FOUNDRY FROM RECYCLED SCRAP FOR LIGHTWEIGHT PROTECTION APPLICATIONS

Marc Pepi[a]
Brandon McWilliams[a]
Andelle Kudzal[a]
Jian Yu[a]
Matthew Charles[b]
Andrew LaTour[b]
Chris Eonta[b]
Danielle Cote[c]
Sharon Park[d]

[a]-Combat Capabilities Development Command - Army Research Laboratory, Aberdeen Proving Ground, MD
[b]-MolyWorks Materials Corporation, Los Gatos, CA
[c]-Worcester Polytechnic Institute, Worcester, MA
[d]-University of Central Florida, Orlando, FL

Abstract
US Army Directive 2019-29 ("Enabling Readiness and Modernization through Advanced Manufacturing") envisions advanced manufacturing transforming battlefield logistics through on-demand fabrication of parts close to the point of need [1]. Towards this end goal, the Combat Capabilities Development Center – Army Research Laboratory (CCDC-ARL) is researching technologies that will make it possible for the warfighter to manufacture and/or repair components at the point of need with scrap materials on-hand, thus increasing operational readiness while reducing the military logistics footprint. The objective of this project was to show a proof-of-concept of additively manufacturing metal powder that was gas atomized in a Mobile Foundry housed within a shipping container from scrap rolled homogeneous armor (RHA) steel. The resultant alloy powder was subjected to laser powder bed fusion (LPBF), for builds that were tested metallurgically and ballistically. Cursory results showed that the additive manufactured builds from powder made from recycled RHA had a martensitic structure as stress-relieved, exhibited excellent mechanical properties, and improved ballistic properties versus this alloy in the wrought condition.

Introduction
One of the many problems encountered with taking part in overseas warfare, is the logistics involved with transporting materiel to the point of need. In the context of this paper, this includes the replenishment of weapons system repair and/or replacement parts. Although Army units currently carry some parts on hand, it is impossible to know exactly which part is going to fail, and surprises will surely be encountered. In some cases, a part failure may render the weapon system idled until the failed part can be replaced or repaired. In some cases, parts can have a long lead time, and not arrive for months. In a worst case scenario, manufacturers may no longer exist for parts on our aging legacy systems. This is the scenario where point-of-need manufacturing

would make a lot of sense – as a means to augment our current supply chain (not replace it) to get our systems up-and-running until the "real" part arrives. The vision set forth in this research is the ability to gas atomize metal powder from recycled scrap in an ISO container for subsequent additive manufacturing. This would align with the Army's number one priority: "Army readiness". Research and development is well underway towards production of metal powder in an ISO container; metal additive manufacturing (AM) at the point of need is still a ways off as of this writing.

Objective / Motivation

As stated in *Logistics Transformation – Reducing the Logistics Footprint* by Ransom et al., "Technology will be one of the primary enablers to reduce the logistic footprint, and the reduction of the logistic footprint is clearly a key element of the future battlefield" [2]. Further, Silva and Rezende note, "The implications of additive manufacturing for logistics could be massive, both for the upstream supply chain and for downstream customer" [3]. In addition, Manners-Bell and Lyon have surmised that additive manufacturing will have implications on logistics by allowing for "near-source" production [4]. To address the role of advanced manufacturing on operational readiness, US Army Directive 2019-29 ("Enabling Readiness and Modernization through Advanced Manufacturing") was published in September, 2019. The objectives of this directive include: "on-demand fabrication of parts close to the point-of-need, and the use of advanced manufacturing at the tactical level to reduce the logistics footprint".

As pointed out by Dellarocco, "shrinking the logistics tail is an important benefit of utilizing indigenous materials in-theater" [5]. To this end, the vision of the research performed herein is to utilize scrap steel armor on the battlefield as feedstock for gas atomization in an ISO container for the purpose of additively manufacturing lightweight protection applications close to the point of need. This paper summarizes the cursory results of this undertaking, from a laboratory setting.

Steel Recycling

It is well known that recycling provides material savings because of a reduction in the need for mined raw materials, and for most metals, recycling can be performed infinitely, in principle [6]. In the commercial sector, the steel circular economy is very well developed (up to 90% recycling rate) [7, 8, 9, 10, 11], providing confidence that successful steel recycling could also be performed at the point-of-need for the military. Scrap steel would most likely be prevalent at a forward operating base (FOB), or contingency operating base (COB) close to the point of need. A potential challenge, however, is associated with contamination of recycled steel (and a resultant reduction in mechanical properties), as residual elements become more abundant with subsequent recycling [6, 8, 11, 12, 13]. The chemical composition of the recycled steel will need to be monitored and controlled.

Scrap Rolled Homogeneous Armor

Rolled homogeneous armor (RHA) is used extensively for ballistic resistance [14], and is the principal material employed for heavy combat and recovery vehicles [15]. Scrap RHA (shown in Figures 1a and 1b), was sent from CCDC-ARL to MolyWorks Materials Corporation for melting and gas atomization processing within the Mobile Foundry. This material was processed as described in the following section.

Figure 1 Two batches of scrap rolled homogeneous armor shipped to MolyWorks for melting and gas atomization, pre- (a) and post-grit blasting (b) to remove corrosion products prior to melting and atomization.

<u>Mobile Foundry Gas Atomization</u>

There are a few technologies that have been shown to successfully additive manufacture using waste materials (or produce AM-Grade powder from scrap), with varying potential to operate in an expeditionary setting. Samarjy and Kaplan have studied laser cutting as a source of molten droplets for AM, and was able to demonstrate a continuous track of waste steel on a steel substrate [16]. Fullenwider et. al. have used a two-stage ball mill to create AM-Grade powder from recycled machining chips [17]. The processed powder was reportedly near-spherical and exhibited a particle size range of 38-150 μm, and was successfully deposited using Laser Engineered Net Shape (LENS) deposition. Likewise, the MELD additive friction stir (AFS) process and cold gas dynamic spray technologies have also shown proof-of-concept of being able to deposit powder made from recycled materials [18]. The MELD AFS process has also been used to deposit material from waste machining aluminum chips [19]. Speaking of machining chips, Mahmood et. al. has reported AM (through laser direct metal deposition) using carbon steel machining chips which were waste from a CNC process [20]. Similarly, Jackson et. al. used the same AM technology to build with 316L stainless steel machining chips [21]. All of these examples show that this concept is of interest across many diverse areas of discipline.

In light of the vulnerabilities presented by convoys restocking outposts with vital supplies, it was determined that the capability of being able to gas atomize scrap metal in an ISO container to produce AM-Grade metal powder at the point of need was a better option than having metal powder shipped via traditional logistics supply channels. This would meet one of the challenges facing metal AM at the point of need. CCDC-ARL awarded MolyWorks Materials Corporation SBIR funding to research and develop a means to accomplish this.

A total of 180 pounds of scrap RHA was sent to MolyWorks Corporation for conversion to AM-Grade powder in their Mobile Foundry. After grit blasting, the panels were melted and atomized to AM-grade powder in the 0-45μm and 45-150μm particle size ranges in the Mobile Foundry. The 0-45μm particle size distribution is useful for laser powder bed fusion (LPBF), cold spray,

MELD, and other AM processes, while the 45-150μm distribution is typical of the range required for LENS deposition. The following sections summarize the results of characterization performed on the resultant powder.

<u>Powder Characterization</u>

The resultant metal powder was subjected to characterization in the form of nanoindentation hardness, metallography, particle size distribution, scanning electron microscopy and chemical analysis.

Nanoindentation Hardness Testing

Nanohardness testing was performed on the powder from scrap RHA by Worcester Polytechnic Institute (WPI). The results of nanohardness testing of the powder from scrap RHA (as well as other metal powders gas atomized by MolyWorks) are listed in Figure 2 [22]. The equipment utilized was the Keysight G200 Nanoindentation Testing System Equipped with a Berkovich diamond indenter tip machined by Micro Star Technologies. The bubble diagram of elastic modulus vs. hardness for this powder had the most error of any of the powders analyzed. RHA "Revert" refers to powder made from RHA scrap that was not classified during the initial run in the Mobile Foundry, and subjected to re-melt and reprocessing. It was interesting to note that the hardness of the 0-45μm batches of RHA exhibited higher hardness than the revert batches.

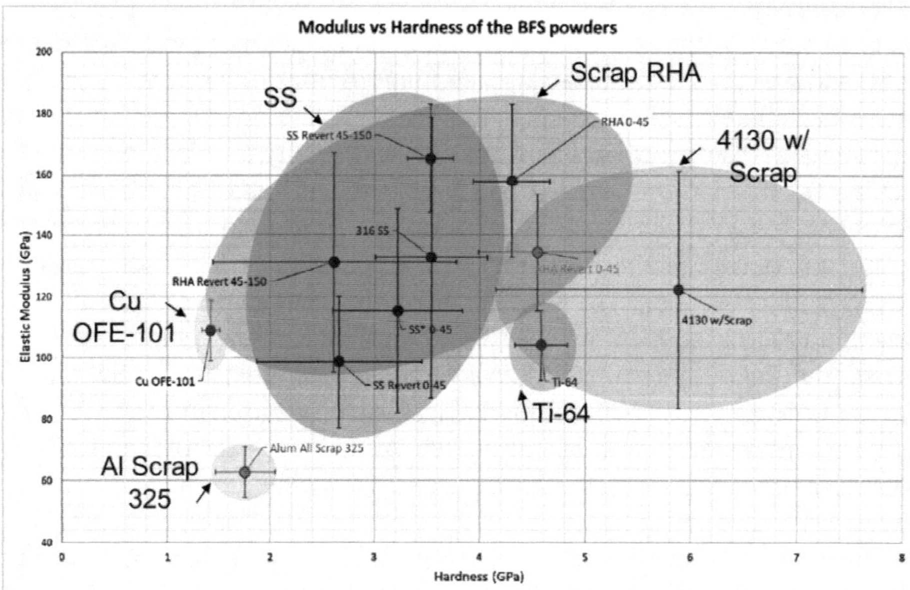

Figure 2 Results of elastic modulus versus hardness (as determined through nanoindentation testing) for various powders produced by MolyWorks. Scrap RHA bubble plot is colored purple.

Metallography

Optical microscopy was utilized to characterize the powder made from scrap RHA. The as-received powder was compression mounted using a Beuhler Simplimet 4000 Mounting System in phenolic resin known as red PhenoCure and polished in a multistage process to produce a mirror finish without compromising the powders integrity. The mirror finish was obtained using a 0.05

μm colloidal silica suspension and cleaned using ultrasonic vibration in ethanol. The polished samples were subsequently examined using the Olympus GX53 Inverted optical microscope. Some non-spherical particles are noted in these photomicrographs (see Figures 3 and 4). Similar trends were noted for the "revert" RHA powder.

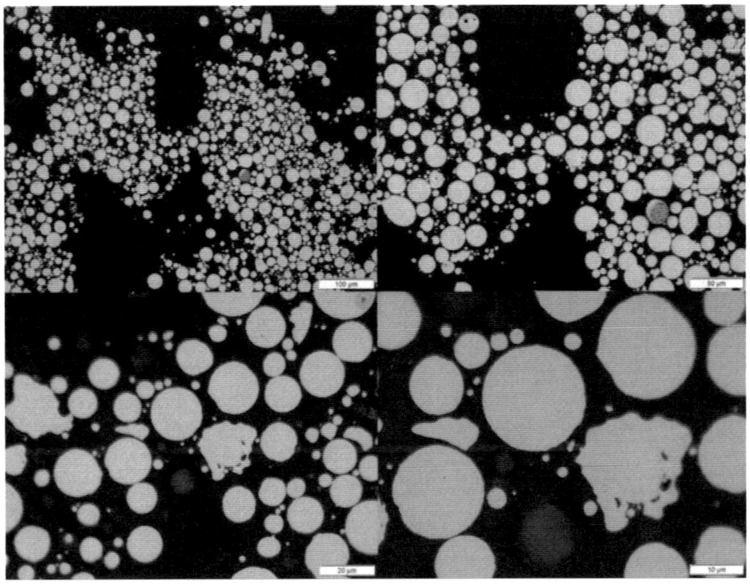

Figure 3 Optical cross sections of the 0-45 μm RHA powder.

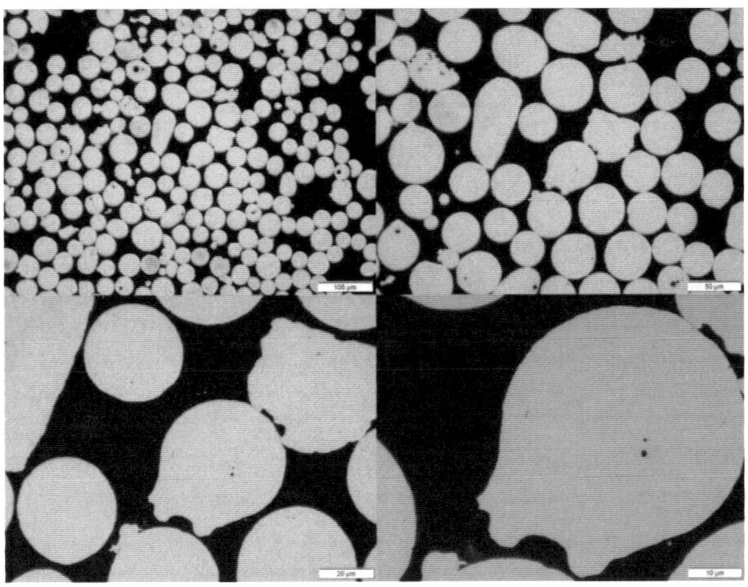

Figure 4 Optical cross sections of the 45-150 μm RHA powder.

Particle Size Distribution

The particle size distribution of the RHA of both size ranges was calculated by the University of Central Florida (UCF) using the Beckman Coulter LS 13 320 with 7 grams of the material. The results are displayed in Figure 5. Each of the results showed that the particle size matched the respective designations for this powder.

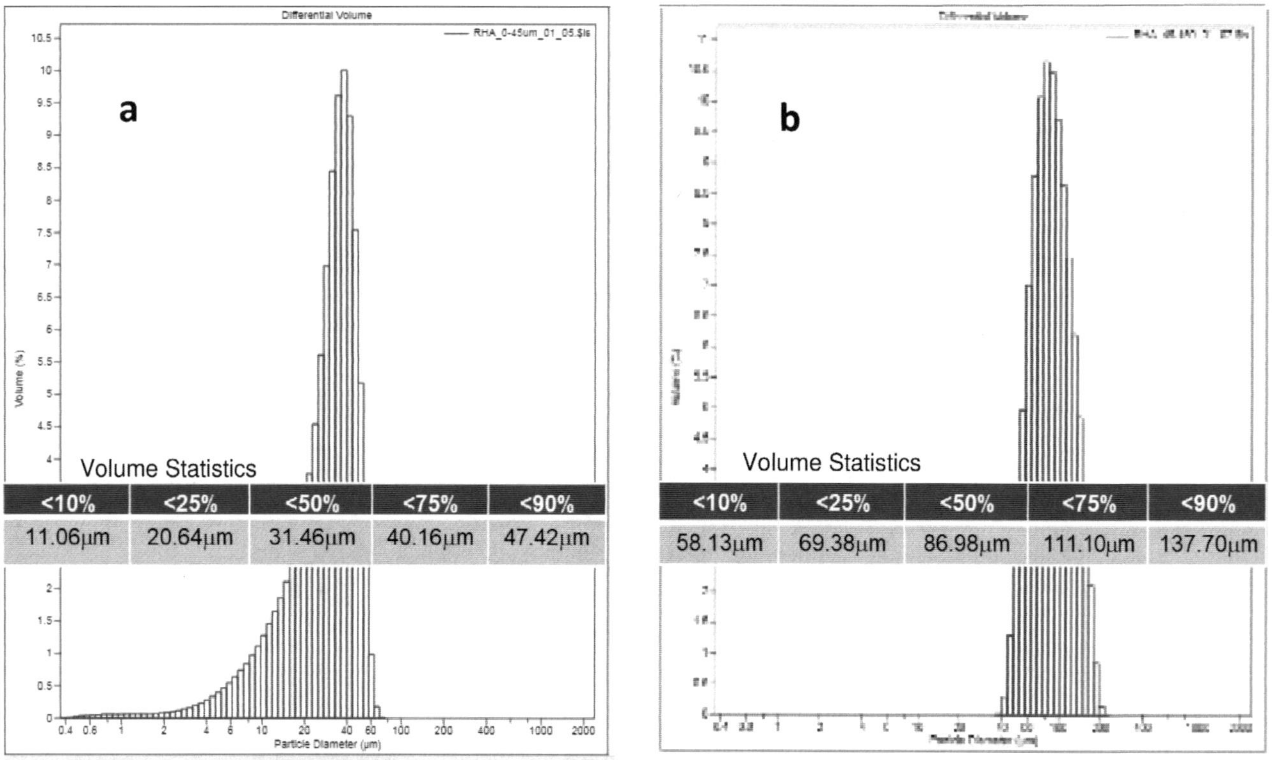

Figure 5 Particle size distributions of the RHA 0-45 µm powder (a) and 45-150 µm powder (b). Particle size conformed to the powder designations.

Scanning Electron Microscopy

The SEM used by WPI for this study was a Zeiss Evo Series Scanning Electron Microscope. The particles were loosely sprinkled upon double-sided SEM mounting tape. To determine homogeneity in terms of geometric particle distribution, SEM imaging was conducted by mounting "loose" powder to a double-sided carbon adhesive stub. To increase imaging quality, the specimens were sputter-coated with gold and palladium particulates. SEM images of the 0-45 mm and the 45-150 mm RHA powder are shown in Figures 6 and 7. Similar to the optical micrographs presented earlier, many non-spherical particles were noted. Some satellites were also noted, a by-product of the gas atomization process.

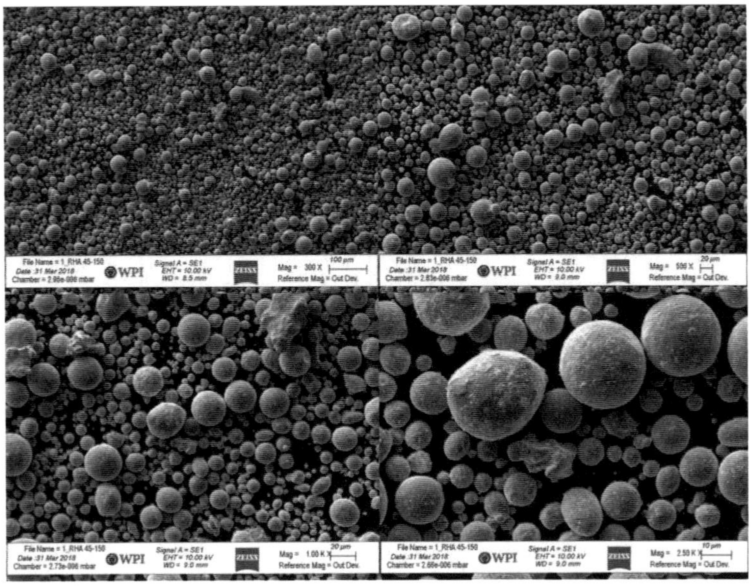

Figure 6 SEM images of the 0-45 μm RHA powder.

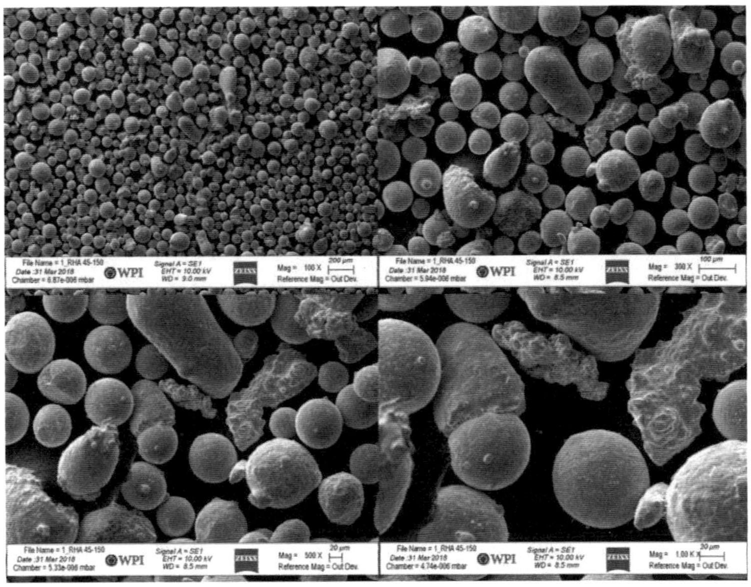

Figure 7 SEM images of the 45-150 μm RHA powder.

Chemical Analysis

Samples of all four powders made from RHA scrap were subjected to chemical analysis by an accredited external laboratory (Luvak Laboratories). The carbon and sulfur were determined through the inert gas fusion technique per ASTM E 1019, while the remainder of the elements were determined using direct current plasma emission spectroscopy per ASTM E 1097. The results are listed in Table 1. The 0-45 μm RHA powder contained titanium, which may have been the result of Mobile Foundry contamination. It also appears the carbon and chromium contents were slightly lower in the revert powder. "Revert" refers to powder that was gas atomized with scrap from the first run.

Table 1
Results of Chemical Analysis
Weight Percent

	RHA 0-45 μm	**RHA 45-150 μm**	**RHA 0-45 μm (Revert)***	**RHA 45-150 μm (Revert)***
Iron	97.8	98.0	98.1	98.1
Nickel	0.029	0.028	0.028	0.027
Chromium	0.32	0.32	0.27	0.27
Manganese	1.05	1.03	1.00	1.02
Carbon	0.167	0.170	0.154	0.151
Molybdenum	0.13	0.11	0.12	0.11
Silicon	0.27	0.26	0.26	0.27
Sulfur	0.005	0.004	0.004	0.005
Phosphorus	0.019	0.018	0.018	0.017
Titanium	0.12	---	---	---

*"Revert" represents a re-run to gas atomize powder from the scrap of the first run.

Additive Manufacturing using Recycled Metal Powder

Much documentation exists with respect to the reuse of metal powder for additive manufacturing [23, 24, 25, 26, 27]. Of course, this is advantageous because of the material and subsequent financial savings. These articles detail the procedures of powder recovery and characterization, provide cost models and even estimate the number of times metal powder can be reused. It was difficult to find any documentation of research into melting and gas atomizing scrap metal into usable AM-grade powder. The powder made from the scrap RHA was subjected to additive manufacturing at MolyWorks and CCDC-ARL.

Laser Powder Bed Fusion of RHA Powder

MolyWorks showed the powder could be used to print an object on the EOS M100. A widget was created using their default 316L stainless steel parameters. Knowing that the powder could be used for additive manufacturing, CCDC-ARL subsequently printed test specimens on their EOS M100 using varied parameters, but settling on what appeared to be optimized parameters similar to those used on a DoD relevant steel.

Once parameters of the LPBF deposition were optimized, the finalized build was completed (see Figure 8). From this, a total of three tensile specimens were machined, and the buttons used for metallography, X-ray computed tomography and hardness testing.

Figure 8 LPBF AM build plate using scrap RHA powder gas atomized in an ISO shipping container.

Metallography

The build plate was stress relieved by loading it into and oven and ramping up to 600C for 2 hours (followed by air cool) prior to machining the specimens. Buttons were metallographically prepared with a 2% nital etchant and examined. The resultant structure is shown in Figure 9. The refined grain structure shown is likely due to the high heating and cooling rate caused by the additive manufacturing process. Note the highly circular defects shown are likely caused by entrapped gas due to the gas atomization process from the powder formation. The structure consisted of fine lath martensite, which is consistent with the structure of a carbon steel that has been austenitized, quenched and tempered. The rapid cooling created during LPBF produces a similar result.

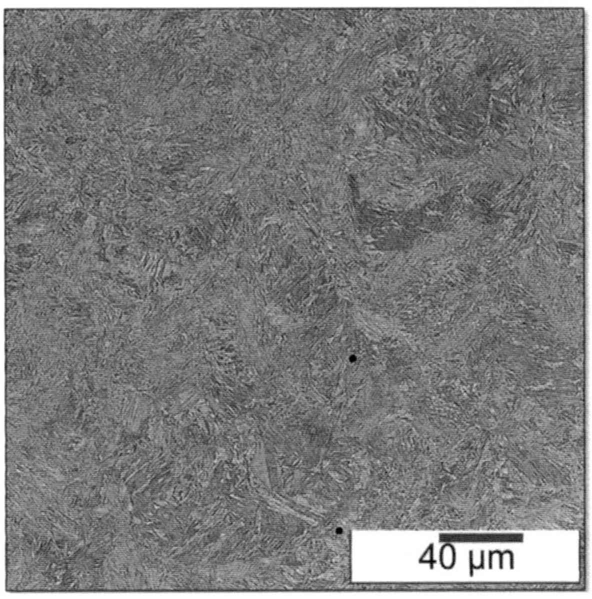

Figure 9 Representative micrograph of the structure of the LPBF AM build using scrap RHA powder gas atomized in an ISO shipping container.

Microhardness Testing
The stress-relieved buttons were subjected to Vickers microhardness testing. The readings were taken on a Wilson Micro Hardness tester using a mapping technique taken at 1mm x 1mm intervals, and the results of 45 readings showed that the material averaged a value of 436±12 HV. This result is actually slightly higher than the average of 423 HV typical of RHA in the wrought condition. This is likely due to the refined grain structured caused by the rapid heating and cooling of the manufacturing process [28] making it comparable to traditionally manufactured RHA without post-processing

Computed Tomography
Cylindrical samples were examined for defects via an x-ray micro CT with a Zeiss Xradia 5200. The porosity was measured using CTan (Bruker) software and the visualization of the defects was generated via CTvox (Bruker) software. The visual representation of the defects (Figure 10) show that they were randomly distributed in the AM part and are generally round in morphology. The AM samples were found to have a density of 99.96%.

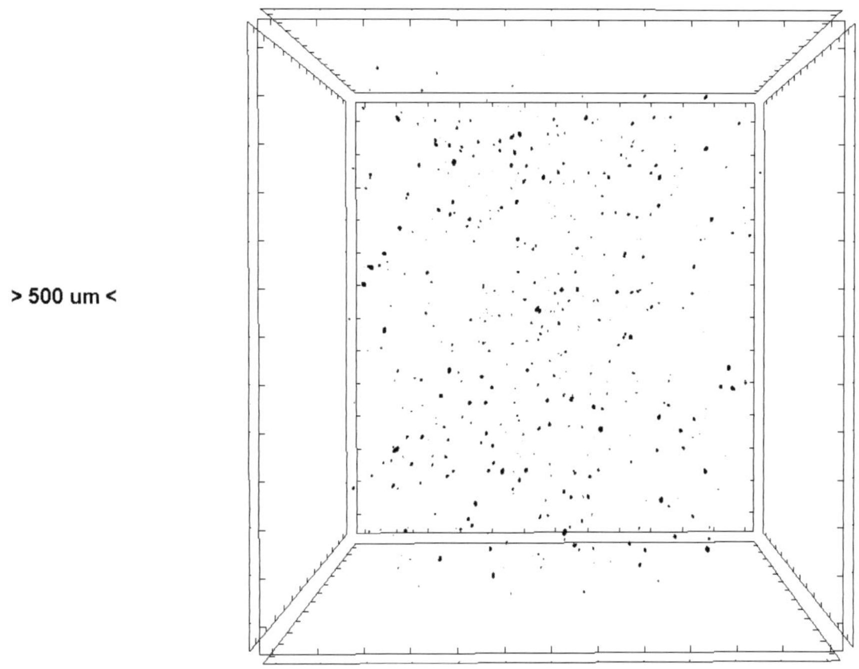

Figure 10 Porosity mapping of a representative section of the LPBF AM build.

Tensile Testing
Three of the nine aforementioned tensile specimens were subjected to testing using the Instron 5985 Series Universal Testing Systems. These three were chosen based on optimized build parameters. The average of the three specimens are listed in Table 2, including the yield strength, ultimate tensile strength, and percent elongation. The mechanical test results listed in Figure 11 show that the samples were repeatable with very similar stress strain curves. The variation of elongation is likely a result of local defects in each sample [29]. Although the UTS of the LPBF

built specimens was lower than that of traditional RHA, the yield strength and percent elongation were comparable. The ultimate tensile strength could likely be improved with a post processing heat treatment.

Table 2
Tensile Test Results

Material	Yield Strength (MPa)	UTS (MPa)	Elongation (%)
AM RHA as Stress-Relieved	1280 ± 25	1320 ± 4	13 ± 2
Traditional RHA [30]	1200	1700	12

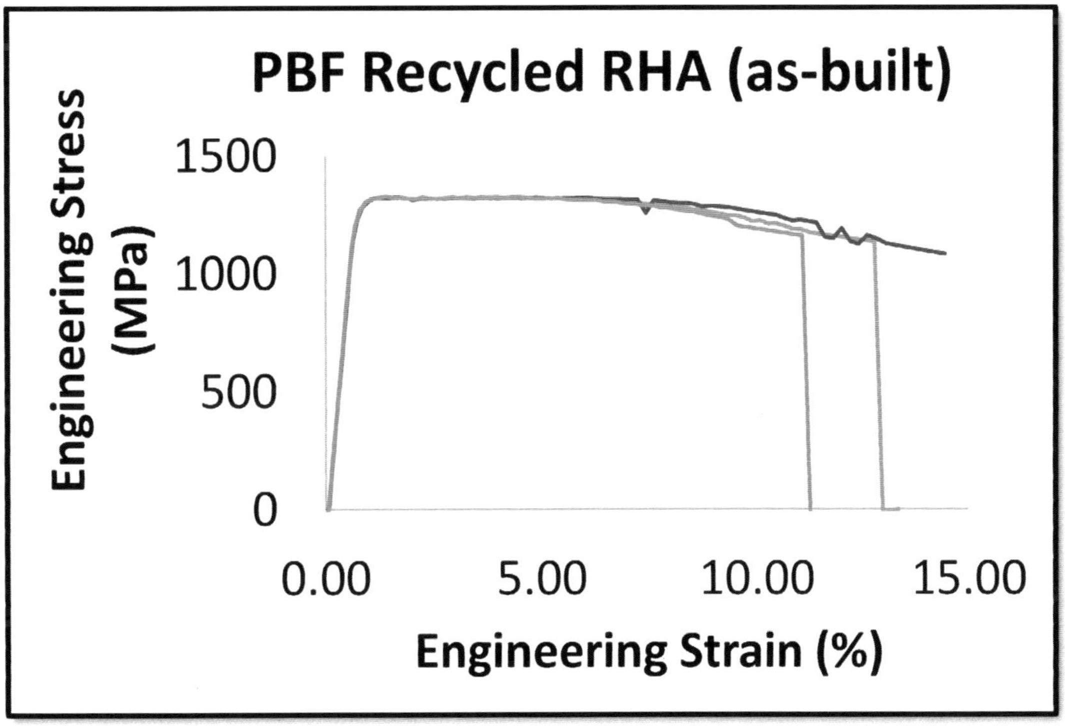

Figure 11 Results of tensile testing of specimens built via LPBF AM using scrap RHA powder gas atomized in an ISO shipping container.

Ballistic Testing
The monolithic plates shown in Figure 12 were used for ballistic testing. Testing was conducted on a gas gun setup described by J. Yu et. al. [31], in accordance with MIL-STD-662F [32]. V_{50} testing was completed using 0.22 caliber fragment simulating projectile (FSP) rounds at 0° obliquity, and the results compared to traditional RHA. Targets were impacted by FSP launched

from a smooth bore gas gun. The target stand was located 0.5m from the gun muzzle. The speed of the projectiles was tracked using a Doppler radar (BR-3502, Infinition Inc.) from the muzzle up to the point of impact. The shot locations on each target were carefully spaced to ensure that the ballistic performance of the target from one impact was not affected by the damage caused by the other neighboring impacts. A 0.02" thick aluminum foil was used as the test witness, which was placed 6 inches behind the target. Any perforation on the witness, by the FSP or any debris from the target, was recorded as complete penetration. Whereas, no perforation was recorded as partial penetration. The V_{50} value was calculated by the average impact speeds of three of the highest partial and the three of the lowest complete penetration impact speeds. At least ten shots were taken for each type of targets. The ballistic properties of plates fabricated via LPBF from recycled RHA powder compared very favorably to the results typical of traditional wrought RHA steel, as shown below (in fact, they showed a 14% increase):

V_{50} (0.22 cal FSP): 440 m/s
V_{50} wrought RHA (0.22 cal FSP): 385 m/s

Figure 12 Ballistic plates built via LPBF AM using scrap RHA powder gas atomized in an ISO shipping container.

Conclusions / Lessons Learned

- Scrap RHA was successfully gas atomized into AM-grade powder in a Mobile Foundry contained within an ISO shipping container.
- The 0-45 mm RHA powder was effectively additively manufactured using the laser powder bed fusion (LPBF) method.
- The resultant structure of the LBPF parts was martensitic, exhibiting high strength.
- AM-built plates made from the powder from scrap RHA achieved a 14% increase in ballistic performance against 0.22 caliber FSP compared to wrought RHA.
- The first ever laser powder bed fusion processing parameters for RHA were developed.

Future Work

Work is ongoing to understand metallurgy of AM metals with respect to high strain rate deformation. Future work may also include further testing to include corrosion and fatigue testing. Various post-heat treatments can also be experimented with to improve the UTS of the LPBF builds.

Acknowledgements

The authors wish to thank Bryer Sousa, Derek Tsaknopoulos, Chris Massar, Kyle Fitzpatrick-Schmidt, Caitlin Walde, Matt Gleason and Jack Grubbs (Worcester Polytechnic Institute) as well as Luvak Laboratories for powder characterization, Josh Taggart-Scarff for metallography and microhardness support, Kevin Doherty (CCDC-ARL) for MELD Manufacturing research and development information.

References

[1] US Army Directive 2019-29, "Enabling Readiness and Modernization through Advanced Manufacturing", dated 18 September 2019.

[2] "Logistics Transformation – Reducing the Logistics Footprint", LTC Darrell S. Ransom, 5 April 2002.

[3] J. Silva et. al., "Additive Manufacturing and its Future Impact in Logistics", 6th IFAC Conference on Management and Control of Production and Logistics, The International Federation of Automatic Control, September 11-13, 2013, Fortaleza, Brazil

[4] J. Manners-Bell and K. Lyon, "The Implications of 3D Printing for the Global Logistics Industry", Transport Intelligence, August 2012.

[5] "Force Projection Research and Development: The Key Enabler for Army Transformation", Col Genaro J. Dellarocco, 10 April 2001.

[6] B. Keck and T. Graedel, "Challenges in Metal Recycling", Science, Volume 337, 10 August 2012.

[7] J. –P. Birat, "Life-cycle assessment, resource efficiency and recycling", Metall. Res. Techno. 112, 206 (2015).

[8] J. Bowyer, et. al., "Understanding Steel Recovery and Recycling Rates and Limitations to Recycling", Dovetail Partners, 3/23/15.

[9] L. Brimacombe, et. al., "Recycling, Reuse, and the Sustainability of Steel", Millennium Steel, 2005.

[10] C. Broadbent, "Steel's Recyclability: Demonstrating the Benefits of Recycling Steel to Achieve a Circular Economy", Springer International Journal of Life Cycle Assessment (2016) 21:1658–1665.

[11] K. Daehn, et. al., "How Will Copper Contamination Constrain Future Global Steel Recycling?", Environmental Science and Technology, ACS, 2017.

[12] H. Ohno, et. al., "Unintentional Flow of Alloying Elements in Steel during Recycling of End-of-Life Vehicles", Journal of Industrial Ecology, 25 January 2014.

[13] C. Marique, "Scrap Recycling and Production of High Quality Steel Grades in Europe", LA Revue de Metallurgie – CIT, November 1996.

[14] W.R. Whittington, et. al., "Capturing the Effect of Temperature, Strain Rate, and Stress Rate on the Plasticity and Fracture of Rolled Homogeneous Armor (RHA) Steel", Materials Science & Engineering A, 594 (2014) 82-88.

[15] J. Prifti et. al., “Improved Rolled Homogeneous Armor (IHRA) Steel through Higher Hardness”, ARL-TR-1347, April, 1997.
[16] R. Samarjy and A. Kaplan, “Using Laser Cutting as a Source of Molten Droplets for Additive Manufacturing: A New Recycling Technique”, Materials & Design Volume 125, 5 July 2017, Pages 76-84.
[17] B. Fullenwider, et. al., “Two-Stage Ball Milling of Recycled Machining Chips to Create an Alternative Feedstock Powder for Metal Additive Manufacturing”, Powder Technology 342 (2019) 562–571.
[18] M. Pepi, et. al., “Progress towards Expeditionary Production of AM-Grade Metallic Powder and Subsequent Additive Manufacturing Applications”, Proceedings of RAPID 2019 conference, Detroit, MI, 5/20/19.
[19] J. Jordon et. al., “Direct Recycling of Machine Chips through a Novel Solid-State Additive Manufacturing Process”, Materials and Design Journal, accepted for publication, and in-press as of this writing.
[20] K. Mahmood, et. al., “Laser Metal Deposition of Steel Components using Machining Waste as Build Material”, Optical Society of America, CLEO Applications and Technology Conference, May 2011, ISBN: 978-1-55752-910-7.
[21] M. Jackson, et. al., “A Comparison of 316L Stainless Steel Parts Manufactured by Directed Energy Deposition using Gas-Atomized and Mechanically-Generated Feedstock”, CIRP Annals - Manufacturing Technology 00 (2020) 1-4,
[22] B. Sousa et. al., “Characterization of Metallic Gas Atomized Powders made from Recycled Battlefield Scrap”, 7/24/18, WPI Technical Report.
[23] B. Sarton et. al., “316L Powder Reuse for Metal Additive Manufacturing”, Solid Freeform Fabrication 2017: Proceedings of the 28th Annual International Solid Freeform Fabrication Symposium.
[24] H.P. Tang et. al., “Effect of Powder Reuse Times on Additive Manufacturing of Ti-6Al-4V by Selective Electron Beam Melting”, JOM, Vol. 67, No. 3, 2015.
[25] M. Barclift et. al., “Cost Modeling and Depreciation for Reused Powder Feedstocks in Powder Bed Fusion Additive Manufacturing”, Solid Freeform Fabrication 2016: Proceedings of the 26th Annual International Solid Freeform Fabrication Symposium.
[26] M. Renderos et. al., “Ni Based Powder Reconditioning and Reuse for LMD Process”, Elsevier, Physics Procedia 83 (2016) 769 – 777.
[27] S. Ford and M. Despeisse, “Additive Manufacturing and Sustainability: An Exploratory Study of the Advantages and Challenges”, Journal of Cleaner Production 137 (2016) 1573 – 1587.
[28] S. Deb Nath et. al., “Microstructure-Property Relationships of 420 Stainless Steel Fabricated by Laser-Powder Bed Fusion”, Powder Technology 343 (2019) 738–746
[29] T. Voisin, et al. "Defects-dictated tensile properties of selective laser melted Ti-6Al-4V." *Materials & Design* 158 (2018): 113-126.
[30] W.N.M. Jamil, et. al., “Mechanical Properties and Microstructures of Steel Panels for Laminated Composites in Armoured Vehicles”, International Journal of Automotive and Mechanical Engineering (IJAME), Volume 13, Issue 3 pp. 3742 - 3753, December 2016.
[31] J. Yu, et. al., “The Current Capabilities on Dynamic Impact Testing”, ARL-TR-4496, July 2008.
[32] MIL-STD-662F, “V50 Ballistic Test for Armor”, 18 DEC 1997.

Resonance Effect, Critical and Resonance Velocities Applied to Diagnostics, Stability and Balancing Methods of Turbine and Generator Rotors over 40 MVA

Zlatan Racic, Marin Racic, A.Yu. Zhyvotov, Yu.G Zhyvotov,

(Original presentation at "13th World Congress in Mechanism and Machine Science,Guanajuato, Mexico, 19-25, June 2011)

Abstract (revised 7-12-2020)

Standard power generation industry practices of inspection, machining, balancing, assembly and bearings and couplings alignment have been developed and streamlined by OEMs for newly manufactured rotors, and implicitly assume that rotors being installed are within design dimensional specifications and tolerances. However, applying these standard practices in the service industry can fall short of expected performance, when dealing with **continuous rotors** with inherent eccentricities outside such specifications. In practice the measured rotors' runouts are equated "unbalances", which excite specific mode at specific operating angular velocity. In this paper we will determine the range of velocities of unsteady rotor's rotation. It is shown that resonance velocities of continuous rotor are not dependent on rotor natural fundamental harmonic resonance frequency, but on the value of accelerating rotating moment that ensures increase of rotor's angular velocity and transfer of rotating and non-rotating reference frames through critical velocity. It is explained also that the critical velocity is not equivalent to harmonic resonance oscillation theory in physics. The specific role of rotor inherent resonance in preparation of rotor to transfer rotating and inertial reference frames through critical velocity range is addressed. It is shown that the resonance phenomenon changes the location of center of mass of the rotor body relative to the journals rotation axis and rotor's geometric center as well as creates conditions for self centering of rotor's body mass axis. The presented research is based on inertial theory of rotor's dynamics based on equation of motion developed impulse and momentum method. The result of this research and understanding of continuous rotor dynamics is of particular importance when developing balancing methods of rotors in power generation industry of T-G set over ~40 MVA of generating electric power, which should be introduced as a separate subsection of the ISO21940-1 standard.

Keywords: Critical Velocity, Natural Resonance, Stability, Reference Frames, Constraints, Balancing

Foreword:

On Resonance, 1st Critical Speed and Stability at Operating Speed of Continuous Rotor with Eccentric Mass Axis in Hydrodynamic Bearings of Large Horizontal Rotating Machine

The focus in this paper is on large turbine and generator rotors which ***operating speed frequency*** *must be less than* ***the natural resonance frequency*** *of rotor inherent natural resonance in free state in gravitational environment, and the fundamental system harmonic mode (1st critical speed frequency) must be less than* ***0.5x*** *of natural harmonic frequency spatial wave. (Fig.2).The allowable residual mass axis eccentricity after balancing should be stated in ISO1940-1 at machine 1st critical.*

All machines designed by man are linear mechanical systems which convert the three fundamental energies in nature: weak interaction, strong interaction and electromagnetic, to external useful work from internal energy of molecular masses of geometric bodies in gravity environment, thereby increasing the work efficiency of direct human labor. Energy is converted via interaction by kinetic energy (motion) or potential energy (stress and strain by pressure and temperature). Machine elements in linear motion must be "rigid" in order to be functional as energy converters and stable to do a useful work at designed operating speed.
When external torque force is used to accelerate, a concentric or balanced rotor in rotating machine, into circular motion from state of rest, rotating machines systems become also energy converters. A torque power (external energy) directs rotor mass to circular motion, and to rotor internal stress and strain from centripetal forces in form of potential energy, with reactive "centrifugal" constant stress forces proportional to angular velocity, acting radially out on particle masses axially distributed as centroidal mass axis between constraints, bound by internal electromagnetic energy interaction on molecular level, with a binding strength based on material atomic properties.

Circular motion:

Energy conversion from external torque to rotor mass circular motion, centripetal acceleration, internal stress and centrifugal forces of continuous rotor with mass axis eccentricites, in horizontal machine, constrained by gravity at bearings constraints at rotor system critical velocity (1st critical speed range) up to ~ 90 degrees phase

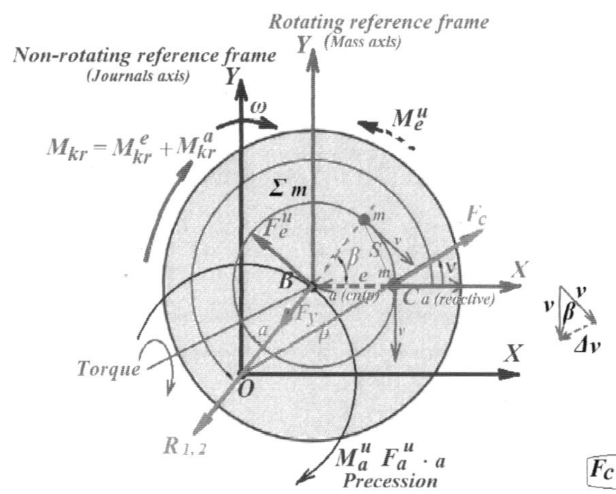

We approximate the arc S bty the cord here to derive the acceleration, but the cord approaches arc for small angles and in thge limit, the result we get is exact.

$$\beta = \frac{S}{e} = \frac{v \Delta t}{e}$$

We can draw a similar triangle with the velocities and conclude:

$$\beta = \frac{\Delta v}{v}$$

Setting the two expressions for β *equal and solving for the acceleration gives:*

$$a \ centripetal = \frac{\Delta v}{\Delta t} = \frac{v^2}{e}$$

$$a \ reactive = - \frac{v^2}{e}$$

$$F_{c\,(centrifugal)} = m \, \rho \, \omega^2$$

For continuous rotors:

$$Fc = ((m(COM\,1) \, x \, e_1) + (m\,(COM2) \, x \, \rho) + (M(COM) \, x \, e_2)) \, x \, \omega^2$$

FW- Figure 1

Rotating machines are efficient energy converters of external energy to external work, designed on the **assumed** equivalency of circular and linear motions and Newton's laws. Continuous rotor, as machine element of finite length, must also be sufficiently "rigid" and stable in system sub- critical, critical and super-critical design operating velocities ranges for a pseudo static stable operation (stable inertial reference frame). With that in mind we must consider that rotor speed and rotating machine energy conversion efficiency, at maximum system external torque, with rotor constrained by gravity in linear oscillating flexible supports of the **appropriate total system stiffness**, must also behave as non-oscillating and "rigid" within machine design operating range and it is dependent on rotor material properties and rigidity (Fig. 10).

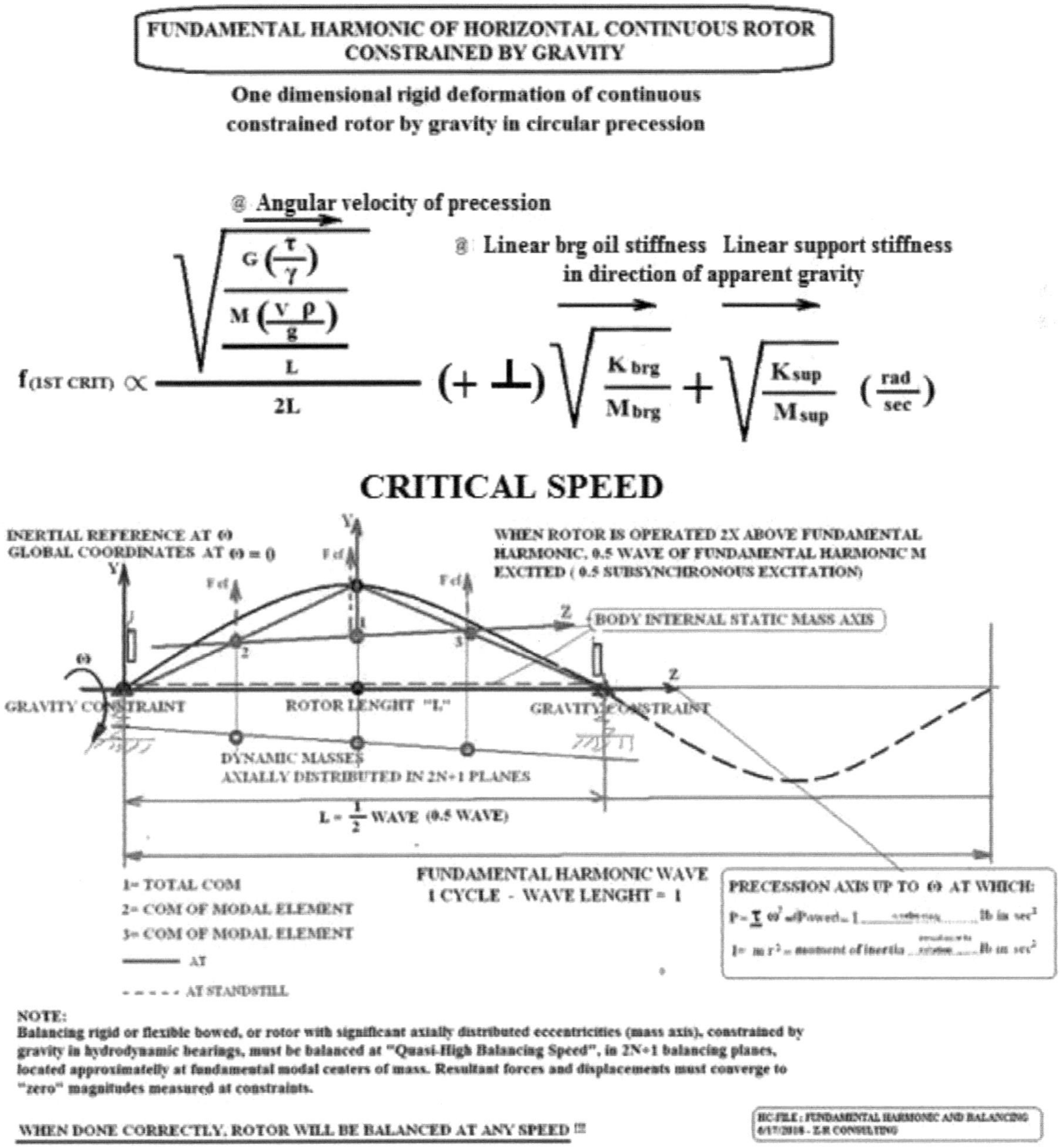

FW-Figure 2

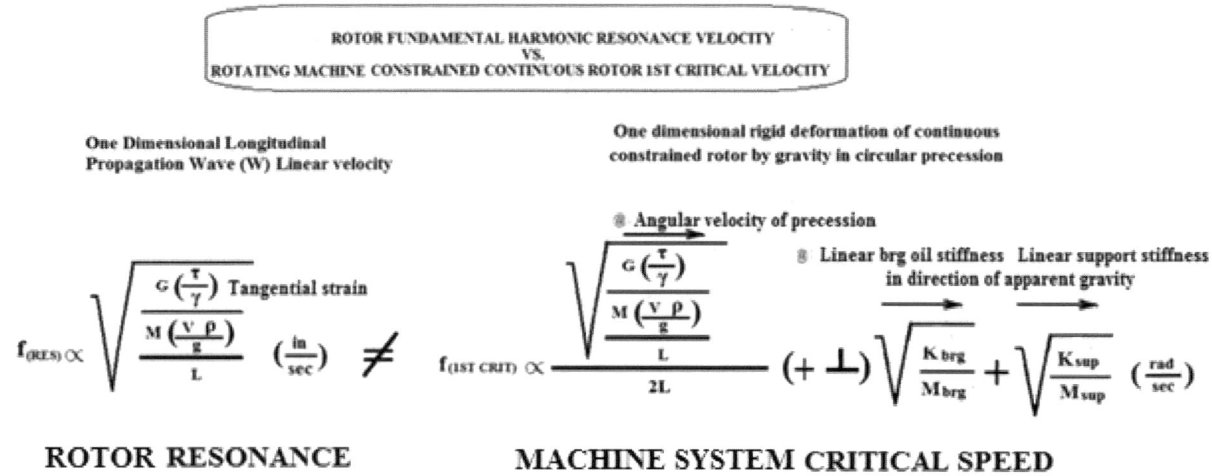

Rotating horizontal machines as energy converters are open systems consisting of a solid body continuous rotor-shaft of finite length with journals constrained in bearings and supports, in non-rotating (inertial) reference frame, with its mass axis referenced as rotating (non-inertial) reference frame up to system "fundamental harmonic" frequency range. The two reference frames of geometric body relative to each other within solid body rotor, are at same time a separate, independent pseudo static stability reference frames of rotor in a system sub-harmonic (at 1st critical speed) and separately in super-harmonic design operating frequency. In rotor's state of rest these frames are referenced to Newtonian global coordinates, and with absolutely concentric rotors the two reference frames are coincident. Operating speed of horizontal rotating machine with rotor constrained by gravity is limited to rotor angular velocities and internal stresses, which do not exceed the longitudinal rigidity of rotor material at 0.5 times system spatial fundamental harmonic wave of rotor natural resonance spatial wave in freestate to operate stably (Fig.2)

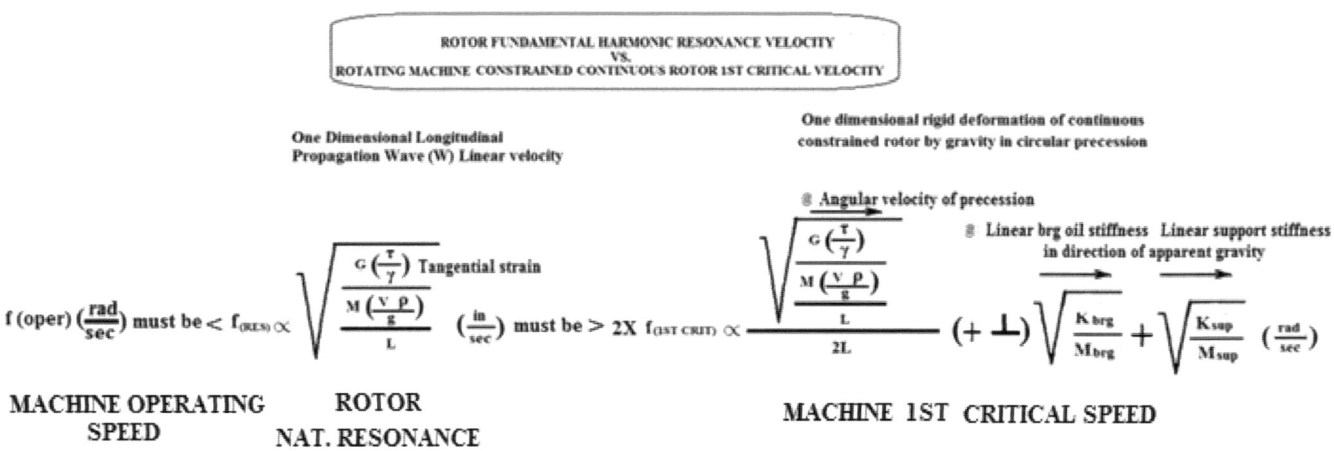

The critical velocity defines the spin frequency, at which by further rotor increase of angular velocity, a forward torque moment overcomes inertia reactive moment of rotor mass lateral moment of inertia in "precession", and rigid mass axis at ~ 90+ degrees phase then becomes a non rotating reference frame centroidal axis, constrained at COM, self centered stable in space, which is different from gravity controlled pseudo static stable state in sub-critical velocity region controlled by gravity and journal/oil friction at point of oil minimum film thickness at sub-critical speeds and at 180 degrees of phase. At super-critical speeds the mass axis is self-centering and observed "pivoting, out of phase" motion is dependent on magnitude of axial asymmetry of the distributed eccentricities of rotor mass between bearings constraints, and the stability threshold speed self centering in space depends on respective stiffness and of each constraint (Appendix 1)

Introduction

Dynamics of a Jeffcott rotor, used here as mathematical model, is the theoretical basis of many-sided interpretations of the continuous rotor dynamic physical events as part of rotating machinery systems. It provides the understanding of physics for the processes that take place during change in rotor velocity in a sub-critical (system sub-harmonic), critical (system harmonic) and super critical (system super-harmonic) velocities. The equations of Jeffcott rotor dynamics were obtained by the methods of classical mechanics using a single disc at COM of the system in rotating and stationary systems of coordinates [1], the Lagrange analytical method of integration in time [2], and also including the Lagrange method to account for the force of external attenuation [3]. For a continuous rotor the same method should be applied to centroidal mass axis as line crossing three points, the COM and two COM s of two fundamental harmonic modal elements in Null mode.

These obtained equations were based on disc on rigid massless shaft, simulating a continuous rotor spinning around geometric center (non-centroidal, non-rotating inertial reference frame), and disc mass axis (centroidal, rotating non-inertial reference frame), laterally precessing around a rotational journals axis in subcritical velocity range [14]. The equations describe the dynamics with disc COM in rotating reference frame is radially offset (eccentric) relative to driving force (torque force) and journals centerline in non-rotating reference frame.

When a concentric rotor is accelerated, the result of conversion from external torque to circular motion of continuous rotor mass is a symmetric axially distributed internal stress in rotor's body between constraints. When rotor mass axis radially deviates relative to journals centerline axis, the radially asymmetric stress from each molecule, axially distributed between constraints create axial moments, radially unidirectional in rotating reference frame between two nodal points. Axially distributed moments disturb relative static equilibrium between adjacent molecules within the mass of geometric body of rotor, causing radial static stress and tangential strain [1] when a flexible eccentric rotor is accelerated through the system critical velocity. The magnitude of stress and linear static deformation of rotor is proportional to rotor spin frequency at angles from ~ 0 degrees to ~ 90 degrees between asymmetric axially distributed stress force, in rotating reference frame, and linear deformation of continuous rotor constrained by gravity at bearings, in non-rotating reference frame. The sum of axially located centrifugal forces in "rigid" rotors is 100% transmitted to journals constraints. The sum of centrifugal forces of axially located centrifugal forces in "flexible" rotors are proportioned to journals constraints and "rigid deformation" up to system critical velocity, and at ~ 90 degrees phase, based on rigidity of rotor material.

In order to simplify visualization of mass axis with infinite number of eccentricities axially distributed in continuous rotor, they can be reduced by superposition to three axial locations on Z-axis: at rotor COM, and two at COM of modal elements of rotor fundamental harmonic Null mode. That provides us with a practical solution for rigid mode balancing by dynamically compensating centroidal mass axis of rotor rigid deformation, by three 180 degrees opposing counter masses, axially proportionally distributed between constraints, at angular velocities up to critical velocity corresponding to phase up to ~ 90 degrees. In case of a torsional twisted eccentric rotor mass axis, correction should be done still in three planes, in orthogonal components.

We conclude that change of location of center of mass takes place instantly after reaching a critical velocity by a rotor in rotating machine at "1st critical speed" at ~90 degrees phase shift. Simultaneously rotor center of mass (COM) becomes a rigid constraint in space (nodal point). Statically deflected rotor body constrained, elastically bent perpendicular to gravity by asymmetric internal radial stress from torque moment between elastic bearings constraints, resume gyroscopic whirling of rotor as two "cantilevers" with journals supported by linear elastic supports and rigidly constrained at COM, resembling the fundamental harmonic mode shape, referenced to inertial (centroidal axis) self-centered in "space" within bearings and seals clearances, in motion of "gull wings", with reduction of observed system amplitudes beyond 90 degrees phase shift, until reaching phase angle of 180 degrees and dynamic and self-centering pseudo-static dynamic equilibrium (self-balancing and self-centered in space in state of least action). If distributed eccentricities are axially asymmetric, with further rotor acceleration at supercritical velocities, the circular motion of journals will be "out of phase" relative to each other as observed by sensor referenced to global coordinates.

Introducing the concept of bearing viscous friction into rotor dynamics [3] allows us to present the resonance phenomenon and rotor reaching pseudo static equilibrium as the process that has specific duration, pseudo statically stable by gravity forces and friction forces between spinning rotor and bearing oil film, at particular angular gravity stability threshold velocity. At the same time the only velocity that defines the resonance phenomenon is the critical velocity of the system. However the experiments show that the resonance phenomenon of rotating machines with flexible continuous rotors starts considerably earlier than the rotation velocity achieves the value of critical velocity (Figure 3). During flexible rotor acceleration from beginning of critical velocity range and a phase shift from 0 to ~ 90 degrees, rotor is undergoing pseudo-static (rigid) deformation [7] due to conversion of external energy to rotor circular motion, and internal radial asymmetric stress and axial strain from constant torque moments (in rotating reference frame), adding to rotor internal potential energy. When rotor in rotating machine reaches critical velocity and phase shift above ~ 90 degrees, and after transfer of reference frames, the stored potential energy is returned to system in form of kinetic energy, until energy is dissipated during further rotor acceleration, and the self- centering effect by inertial forces terminates at the velocities that exceed critical velocity and defines the end of critical velocity range at 180 degrees phase shift, when rotor reaches the state of least action.

The experiment shows (Figures 8 and 9) that resonance phenomena takes place in the definite area of velocities that are close to critical velocity and rotating machine system "1st critical speed". However the limits of this area (stability threshold velocity) were not theoretically determined accurately enough since it is dependent on journal gravity load, bearing bore configuration and oil viscosity. The experiment shows that calculating critical velocity usually does not coincide with the velocity where the maximum rotor shaft peak rotor deflection takes place relative to global (fixed) coordinates. Inertial theory of rotor dynamics [4, 5] showed that preliminary preparation of rotor non-rotating reference frame to transfer through critical velocity starts at sub-critical velocities ω_r. It relates to rotor's mass centroidal axis precessing around journals axis up to the phase angle of ~90° (between internal axial moments, and external rotor geometric deflection). At the same time the axially distributed mass axis eccentricity relative to rotor neutral axis is constant (notwithstanding the fact that elastic rotor body undergoes "rigid deformation" and continuous body centroidal axis (rotating reference frame - Z-axis, laterally precessing about journals centerline axis, constrained by gravity against bearing oil stiffness and support system stiffness, while resembling the fundamental harmonic mode shape between constraints, referenced to global coordinates [8].

Note that the main theoretical research of Jeffcott rotor dynamics was conducted in conditions where "rotor-shaft" system rotates uniformly and the necessary rotating moment is created by the external driving gear. It means that a change in rotor's velocity was considered as inessential. In practice the change of rotor's velocity occurs with rotor's acceleration. At the same time the resonance phenomena becomes apparent at the moment when the rotor makes accelerated or decelerated rotation under the effect of rotating moment of the gear or under the effect of inertia moments during the rotor deceleration. All of this means we had to define an area of rotor velocities that are connected with the system's resonance phenomena. We then had to determine rotor velocities when the resonance phenomena of the system are started and ended. Moreover, we had to explain the physics of resonance phenomenon and mechanism of rotor's reference frames transfer at a system 1st critical velocity.

System of rotor dynamic equations

In this section we explain the system of equations that comprehensively describe the uniform rotation of a Jeffcott rotor in conditions of weightlessness and the absence of any external or internal friction. The system of equations determines the transfer mode of rotation with subcritical velocities along with supercritical rotation mode with supercritical velocities. We address this system of equations permanently while also considering other issues, since any accelerated rotor rotation invariably transfers into uniform rotation.

The scheme of rotation and acting inertial moments is explained in Fig.1.

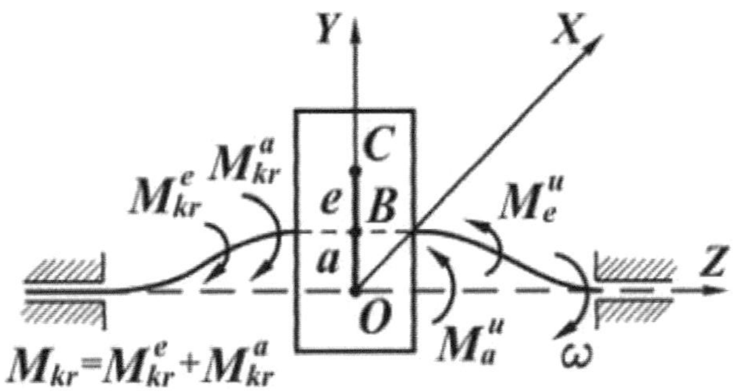

Fig. 1

Fig.1 shows rotation around a shaft's middle line and the shaft's precessing around rotational axis. Also the rotating and inertial moments of a "rotor-shaft" system are noted. Rotating moments act around a shaft's geometric axis. Inertial moments are stipulated by uneven distribution of mass around that geometric axis and rotational axis. Inertial moments simultaneously act around a shaft's geometric axis and around rotational axis. The location of the rotor is shown in Fig.1 and corresponds to rotor rotation with initial velocity equal to the one when (Jeffcott rotor with massless shaft) shaft start bending under the effect of "centrifugal" force (i.e. centripetal reactive force) at ~ 45 degrees phase shift (**β**). (Continuous rotor-shaft undergoes rigid deformation)

The scheme of forces and moments that affect onto rotating rotor is demonstrated on the Fig.2.

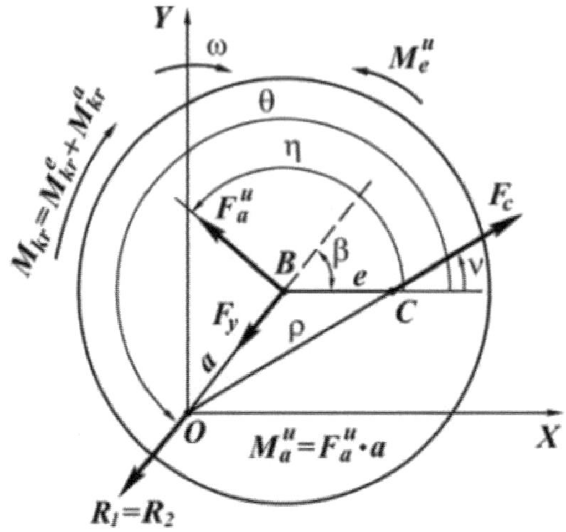

Fig. 2

The system of equations that describes the complete dynamics of a Jeffcott rotor in whole range of velocities looks like this:

$$
\begin{aligned}
& M_{kr} - m\omega^2 e^2 - m\omega^2 a^2 = 0 \\
& M_{kr}^e - m\omega^2 e^2 = 0 \\
& M_{kr}^a - m\omega^2 a^2 = 0 \\
& R_1 - \frac{1}{2} ka = 0 \\
& \sin\beta = \frac{a}{e} \\
& m\omega^2 \left(a + e\cos\beta\right) - ka = 0
\end{aligned}
\qquad (1)
$$

At the same time, the following forces and moments affect "rotor-supports" system with statically unbalanced rotor depending on the rotation velocity:

$$M_{kr} = M_{kr}^{e} + M_{kr}^{a}$$

$$M_{kr}^{e} = m\omega^2 e^2$$

$$M_{kr}^{a} = m\omega^2 a^2$$

$$M_{ob}^{u} = M_{e}^{u} + M_{a}^{u}$$

$$M_{e}^{u} = m\omega^2 e^2$$

$$M_{a}^{u} = m\omega^2 a^2 \qquad (2)$$

$$F_c = m\omega^2 \rho, \qquad \sin\nu = \frac{a}{\rho}\sin\beta$$

$$\rho = \sqrt{e^2 + a^2 + 2ea\cos\beta}$$

$$F_y = ka \qquad \theta = \beta + 180^\circ$$

$$F_a^{u} = m\omega^2 a \qquad \eta = \beta + 90^\circ$$

$$R_1 = R_2 = \frac{1}{2}ka \qquad \theta = \beta + 180^\circ$$

Where:

m ---- rotor mass;

ω — rotor's rotation velocity;

ω_{kr} — critical velocity;

e — rotor eccentricity;

a — rotor shaft bend (deflection);

k — coefficient of rotor elasticity;

β — angle of rotor turn around shaft's middle line;

M_{kr} — total rotating moment that ensures uniform rotor's rotation;

M^{e}_{kr} — rotating moment that ensures uniform rotor's rotation around shaft's journals center line;

M^{a}_{kr} —rotating moment that ensures uniform rotation of shaft middle line together with rotor around rotational axis;

M^{u}_{ob} — total inertial moment that prevents uniform rotation of "rotor-shaft" system;

M^{u}_{e} — inertial moment that prevents uniform rotation of rotor around shaft's middle line;

M^{u}_{a} — inertial moment that prevents uniform rotation of shaft's middle line together with rotor around rotational axis;

F_c — centrifugal force applied towards rotor's center of mass;

ν — angle which is shaped by centrifugal force and direction of vector of imbalance;

F_y — shaft's elasticity force;

θ — angle that is shaped by shaft's elasticity force and direction of vector of imbalance;

$F^u{}_a$ — inertia force that prevents rotation of shaft's middle line together with rotor around rotational axis;

η — angle that is shaped by force and direction of vector of imbalance;

$R_1 = R_2$ — force of supports' reaction;

θ — angle that is shaped by the force of supports' reaction and direction of vector of rotor's imbalance.

This system of equations and scheme of forces and moments can be used to define resonance velocities and a range of steady operative velocities of rotating machine.

Under "resonance velocity" we will consider the velocity that determines the beginning and ending of "resonance range when the rotor velocity is changed.

Resonance velocity of rotor's rotation in supercritical rotation mode

Rotor dynamics usually assumes that the range of operative velocities covers the whole range of supercritical velocities up to the area of auto-oscillations. However the system of equations that describes the uniform rotation of a Jeffcott rotor shows that the range of operative supercritical velocities starts from the velocity that is 1.74 times higher than critical velocity. To explain this let's show some dependencies of relative parameters of rotor rotation from relative velocity on transfer and supercritical mode.

Relative shaft bend from relative velocity on transfer and supercritical mode has the following view:

$$\frac{a}{e} = \frac{\omega^2}{\omega_{kr}^2} \frac{1}{\sqrt{1 - \frac{2\omega^2}{\omega_{kr}^2} + \frac{2\omega^4}{\omega_{kr}^4}}} \qquad (3)$$

$$\frac{a}{e} = \frac{\omega^2}{\omega_{kr}^2} \frac{1}{\frac{\omega^2}{\omega_{kr}^2} - 1} \qquad (4)$$

These dependencies are shown in Figure 3.

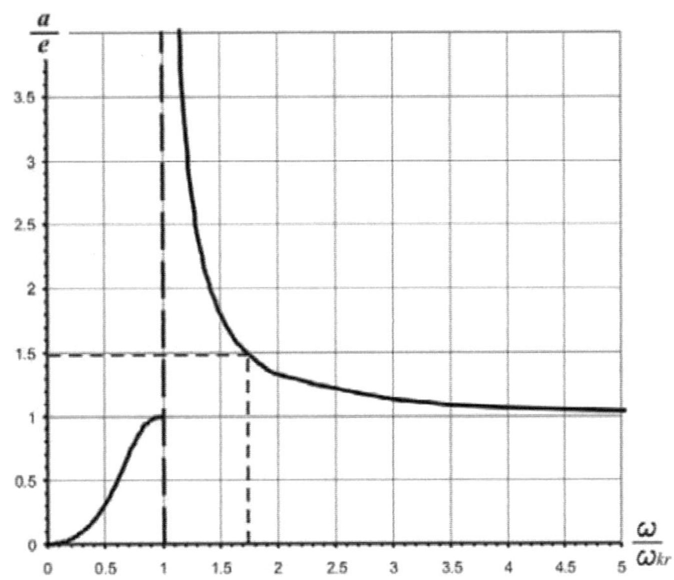

Fig. 3

Let's show the dependency of relative rotating moment M^{e}_{kr} from relative velocity

velocity which has the

$$\frac{M^{e}_{kr}}{ke^2} = \frac{M^{u}_{a}}{ke^2} = \frac{\omega^6}{\omega^6_{kr}} \frac{1}{1 - \frac{2\omega^2}{\omega^2_{kr}} + \frac{2\omega^4}{\omega^4_{kr}}} \qquad (5)$$

$$\frac{M^{e}_{kr}}{ke^2} = \frac{M^{u}_{a}}{ke^2} = \frac{\omega^6}{\omega^6_{kr}} \frac{1}{\left(\frac{\omega^2}{\omega^2_{kr}} - 1\right)^2} \qquad (6)$$

following view:

These dependencies are graphically demonstrated on the Fig 4

Let's pay our attention to the Fig. 4. As it follows from the chart, on supercritical velocities the relative minimal inertial moment

M^{u}_{a} **/ ke =6.75** corresponds to the system's rotation velocity which is approximately **1.74** times higher than critical velocity. This velocity is the minimal velocity $\omega = l,74\ \omega_{kr}$ when uniform rotor's rotation is possible.

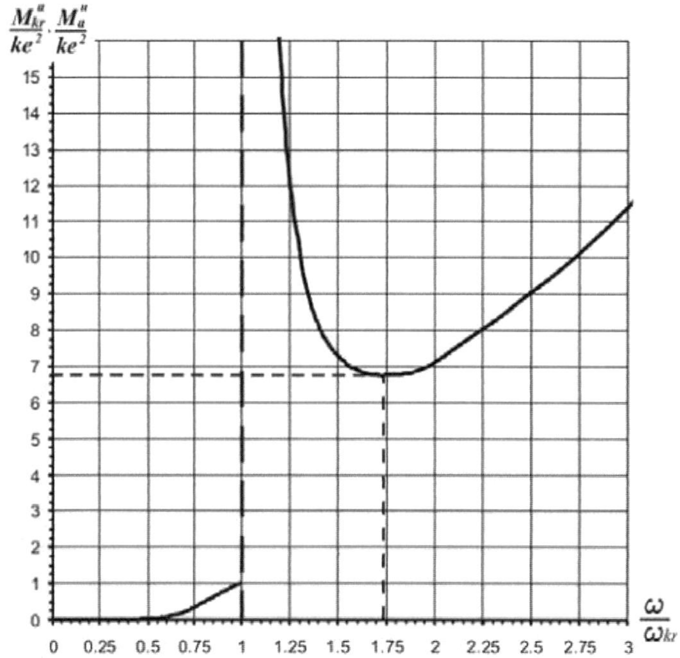

Fig. 4

at supercritical mode of In accordance with the chart (Fig. 3), the minimal velocity corresponds to rotor's rotation when shaft's bend is equal to $a= 1.5e$. Such shaft's bend relates to completion of resonance phenomena on supercritical mode during increasing of rotor's velocity.

If $M^a{}_{kr} / ke^2 > 6.75$, then velocity of rotor's uniform rotation exceeds the minimal velocity $\omega = 1.74.\omega_{kr}$
If $M^a{}_{kr} / ke^2 < 6.75$ then the rotation of the system becomes slower.
Inertial moment (Fig.4) becomes bigger than rotational moment and we monitor the process that is opposite to the process of rotor's self-centering (instability).
Moreover, if relative rotational moment $M^a{}_{kr} / ke^2 < 6.75$ is less than the rotational moment $M^u{}_a / ke^2 = 1$ which corresponds to critical velocity ω_{kr}, then the rotor's velocity reaches one of the velocities of transfer mode.
If $1 < M^a{}_{kr} / ke^2 < 6.75$ then the resonance phenomenon takes place that results in destruction of rotor machine.

Therefore the minimal velocity $\omega = 1.74.\omega_{kr}$ ·, determines the velocity of completion of resonance phenomena. It also determines the range of operative velocities of rotor's uniform rotation up to the moment of auto-oscillations.
At the same time the minimal velocity determines the commencement of resonance

Critical velocity of the steady system rotation

Critical velocity ω_{kr} is substantially considered as the theoretical velocity since it is not possible to confirm it by experiments.
By reaching the critical velocity it creates duality of rotor's rotation parameters. At this velocity the uneven change of shaft's bend, angle of rotor's turn around shaft's middle line, inertial and rotating moment should take place. In mechanics the uneven changes of rotor's rotation parameters are impossible. Any process should be durable (consistent). Absence of resonance's durability (response) is explained by the fact that system of equations (1) of dynamics of Jeffcott rotor was obtained when change of velocity was very slow. The change of velocity could be neglected and we could consider that rotor rotates uniformly. In result all processes required for "rotor- shaft" system's transfer of reference frames through critical velocity is almost instantaneous. **As a result, we had a false impression that critical velocity separates non-linear response across system critical velocity frequency range the transfer of reference frames which occurs on subcritical velocities, from the process of rotor's self-centering which occurs at supercritical velocities, and vice-versa.**

Due to the above-mentioned, it is necessary to consider the rotor dynamics taking into account the affect of additional accelerating moment which ensures uniformly accelerated rotor's rotation. Additional accelerating moment M^a_{yc} can be presented as part of rotating moment M^a_{kr} i.e. we can consider that $M^a_{yc} = 0.1\, M^a_{kr}$ etc.

In result the sum of moments M^a_{yc} and M^a_{kr} can be presented as $\Sigma M^a_{kr} = n\, M^a_{kr}$ where **n>I.**.
The moments M_{kr} and M^e_{kr} will be increased in the same proportion.

System of equations of uniformly accelerated Jeffcott rotor rotation

Taking into account the above comments, let's consider the scheme of forces and moments which affect onto rotor during uniformly accelerated rotation which is demonstrated in Figure 5.
Let's introduce to the scheme the accelerating moment M^a_{yc} which ensures uniformly accelerated rotor's rotation.
The affect of accelerating moment M^a_{yc} which acts onto rotor will result in appearance of additional inertial moment M^a_{yc} which. This inertial moment can be presented as force F^u_{yc} that is acting on the arm a.
Let's apply the D'Alembert principle in order to ~~push~~ transform or the task of dynamics to the task of statics, i.e. let's push the task of uniformly accelerated rotation to the task of uniform rotation of "rotor-shaft" system. At the same time at the scheme presented on the Fig.5 let's demonstrate the rotational moment M^a_{kr} which ensures uniform rotation of rotor as the pair of forces F^a_{kr} which have arm e. The accelerating moment M^a_{yc} is also presented as the pair of forces F^a_{yc} which have arm e.

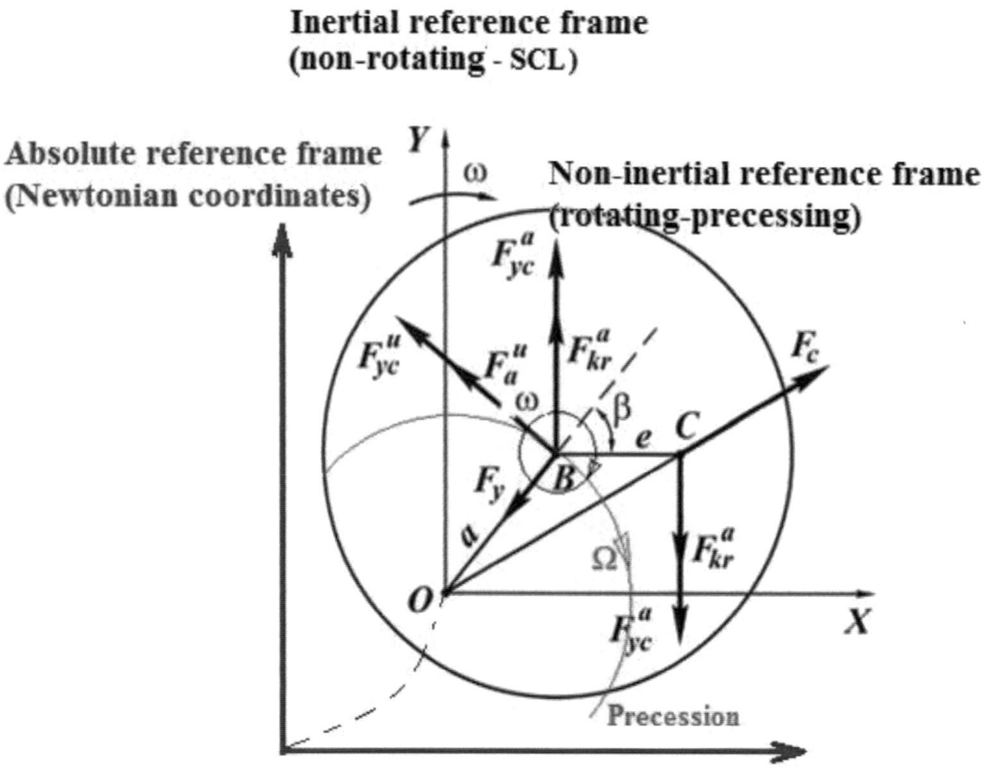

Fig. 5 Disc of a Continuous rotor reference frames

In order to obtain the system of equations we take into consideration the condition of rotor's dynamic balance during uniformly accelerated rotation. In the beginning let 's compile the equation of moments relatively to the point $\boldsymbol{O}$ at definite instant $\boldsymbol{\omega}$.

$$F^u_a + F^u_{yc} - F^a_{kr} - F^a_{yc} = 0 \qquad (7)$$

or

$$nM^u_a = nM^a_{kr} \qquad (8)$$

Since $\boldsymbol{M^u{}_{yc} = n\, m\, \omega^2\, a - m\, \omega^2\, a}$, then it follows from dependency that:

$$F^a_{yc} = nm\omega^2 a - m\omega^2 a = m\omega^2 a(n-1) \qquad (9)$$

Therefore there were defined all forces and moments which affect onto rotor that rotates in uniformly accelerated way around rotational axis.

Let's compile the equation of moments relatively to the point $\boldsymbol{B}$ at definite instant velocity $\boldsymbol{\omega}$.

$$F_c \frac{ea \sin \beta}{\rho} - nM_{kr}^{a} = 0 \qquad (10)$$

or

$$m\omega^2 \rho \frac{ea \sin \beta}{\rho} - nm\omega^2 a^2 = 0 \qquad (11)$$

It follows from dependency that:

$$\sin \beta = n \frac{a}{e} \qquad (12)$$

The dependency (12) shows that rotor turns around shaft's middle line on angle 90° while value of ***a*** is less than ***e*** since:

$$a = \frac{e}{n} \qquad (13)$$

Let's compile the equation of moments relatively to the point ***C*** and definite instant velocity ***ω***:

$$nM_{kr}^{a} + F_{a}^{u} e \cos \beta + F_{ye}^{u} e \cos \beta - F_{y} e \sin \beta = 0 \qquad (14)$$

or

$$nm\omega^2 a^2 + nm\omega^2 ae \cos \beta - kaen \frac{a}{e} = 0 \qquad (15)$$

The final view of Equation 15 will be:

$$m\omega^2 \left(a + e \cos \beta\right) - ka = 0 \qquad (16)$$

For transfer mode of rotor's rotation around rotational axis the equation (16) can be presented as:

$$m\omega^2 \left(a + \sqrt{e^2 - n^2 a^2}\right) - ka = 0 \qquad (17)$$

The equation (17) allows to determine the value of shaft's bend depending on rotor's rotation velocity:

$$a = \frac{m\omega^2 e}{\sqrt{n^2 + k - 2km\omega^2 + m^2\omega^4}} \qquad (18)$$

The equation (17) also allows us to determine the value of angular velocity when the condition of dynamic balance is ensured:

$$\omega = \sqrt{\frac{ka}{m\left(a+\sqrt{e^{2}-n^{2}a^{2}}\right)}} \qquad (19)$$

As it follows from the dependency, the critical velocity of rotor's rotation $\boldsymbol{\omega_{kr}}$ is achieved in case if:

$$e^2 - n^2a^2 = 0, \text{ i.e. } a = \frac{e}{n}.$$

Let us refer to graph of dependency of relative shaft's bend on relative velocity shown in Figure 3. The graph indicates the velocity of rotor's rotation when the resonance phenomenon occurs. Let's call it as resonance velocity ω_r on transfer rotation mode. Since $a/e = 1/n$, then resonance velocity ω_r is always less than critical velocity ω_{kr}..
On the Fig.6 the dependency of relative resonance velocity on value n is presented. It was compiled applying graph on the Fig. 3.

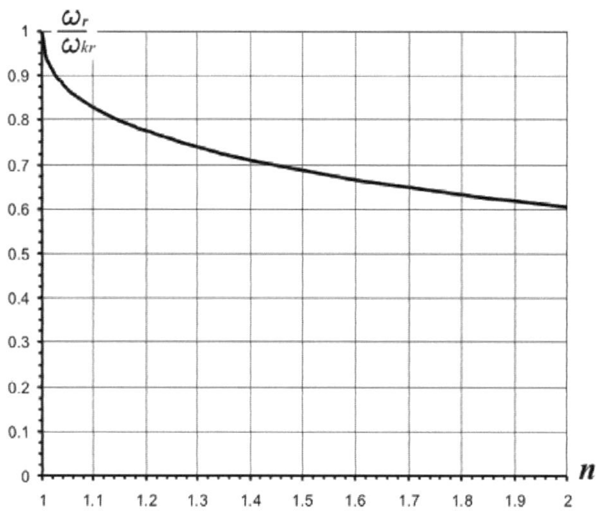

Fig. 6

The Figure 7 presents dependency of rotor's turn angle $\boldsymbol{\beta}$ around shaft's middle line on relative resonance velocity.
This dependency shows that angle $\boldsymbol{\beta}$ is less than 90° for any resonance velocity $\boldsymbol{\omega_r}$ at the moment of resonance phenomena commencement. Therefore, in order to ensure conditions for rotor's self-centering at preset resonance velocity $\boldsymbol{\omega_r}$ then rotor should be rotated beyond an angle 180°- $\boldsymbol{\beta}$.
At the same time the angle $\boldsymbol{\beta}$ should correspond to preset resonance velocity.
The conducted research shows that accelerating moment determines the resonance velocity $\boldsymbol{\omega_r}$ and angle of rotor's turn around shaft's middle line at the moment of resonance

phenomena commencement.
In whole, the conducted research defines the range of velocities which is connected with resonance phenomenon. This range of velocities lies between resonance velocity ω_r of transfer rotation mode and resonance velocity $\omega^S_r = 1.74\ \omega_{kr}$ of supercritical rotor's rotation mode.
Finally, the conducted research explains the physics of resonance phenomenon and effect of rotor's self- centering.

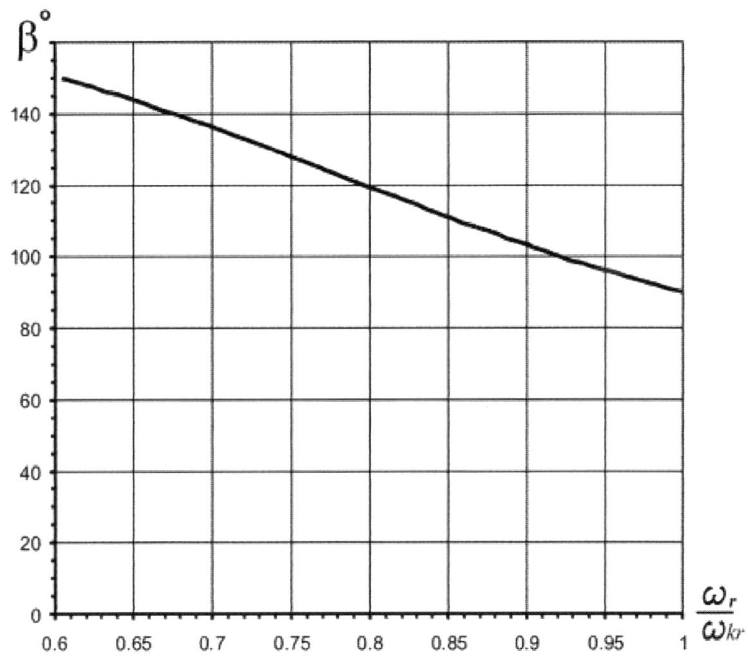

Fig. 7

Physics of resonance effect

In the theory of rotor dynamics the resonance is considered as the phenomenon which is similar to resonance phenomenon in theory of oscillations. Appearance of resonance is explained by coincidence of rotational velocity of "rotor-shaft" system with critical velocity. However the resonance in rotor dynamics considerably differs from resonance of linear oscillation processes. This is the resonance phenomenon that creates conditions for final rotor's turn around shaft's middle line when rotor's center of mass is located between geometrical axis and rotation axis. In result rotor and shaft which were rotating around rotational axis with their heavy side radially outwards., after resonance they rotate with their lightweight side radially outwards..

In order to explain the peculiarities of resonance in rotor dynamics examine Figure 5 with a reference to the action of moments relatively to rotor's center of mass.

The moments that were generated by the pairs of forces $F^a{}_{kr}$ and $F^a{}_{yc}$, strive for overcoming the shaft's elasticity force moment $\boldsymbol{F_y}$ that is defined relatively to the rotor 's center of mass. However, until the projection of forces $\boldsymbol{F^a_{kr}}$ and $\boldsymbol{F^a_{yc}}$, on the line of action of shaft's elasticity force $\boldsymbol{F_y}$, is equal to elasticity force, the location of rotor relatively to the rotational axis will be sufficiently specific. At the same time the rotor is turned relatively to the shaft's middle line ~~on~~ the angle of $\boldsymbol{\beta = 90°}$.

As far as the rotor's velocity exceeds resonance velocity $\boldsymbol{\omega_r}$, the value of forces' projections $\boldsymbol{F^a_{kr}}$ and $\boldsymbol{F^a_{yc}}$, starts to exceed the value of shaft's elasticity force $\boldsymbol{F_y}$. The elasticity force is unable to retain rotor relatively to the rotational axis. At the same time the increase of rotor's velocity assists in unlimited moving-away of rotor from rotational axis. The center of mass and point of rotor's mounting to the shaft are moving radially away from rotational axis. This phenomenon takes place until the moment when rotor turns by an angle of $\boldsymbol{\beta = 90°}$ around shaft's middle line.

In order to understand the reason of rotor's tum, let's consider the affect of moments relatively to the geometric center of rotor (**point *B***). The moments generated by the pairs of forces $\boldsymbol{F^a_{kr}}$ and $\boldsymbol{F^a_{yc}}$, are striving to turn rotor around shaft's middle line overcoming the moment that is generated by centrifugal force around **point *B*.** At rotor's velocity that exceeds resonance velocity $\boldsymbol{\omega_r}$ the moment generated by centrifugal force is not able to counteract to the moments generated by forces $\boldsymbol{F^a_{kr}}$ and $\boldsymbol{F^a_{yc}}$,. Therefore rotor continues to turn around shaft's middle line. At sufficient power of driving gear the alignment of center of mass with a line of shaft's elasticity force action takes place. New location of rotor's center of mass creates conditions for rotor's self-centering (Threshold of stability). At same time, the rotor starts its towards the rotational axis (descending amplitude of journals centers with COM as a rigid constraint in space). Rotor's self-centering occurs, reaching a state of least action and static and dynamic equilibrium. . If the power of driving gear is not sufficient then resonance phenomena result in unstable mode of rotor's rotation and rotor and machine destruction of rotor machine.

The peculiarities of rotor dynamics during transfer through critical velocity are well demonstrated by the results of experiments presented on the Fig.8.

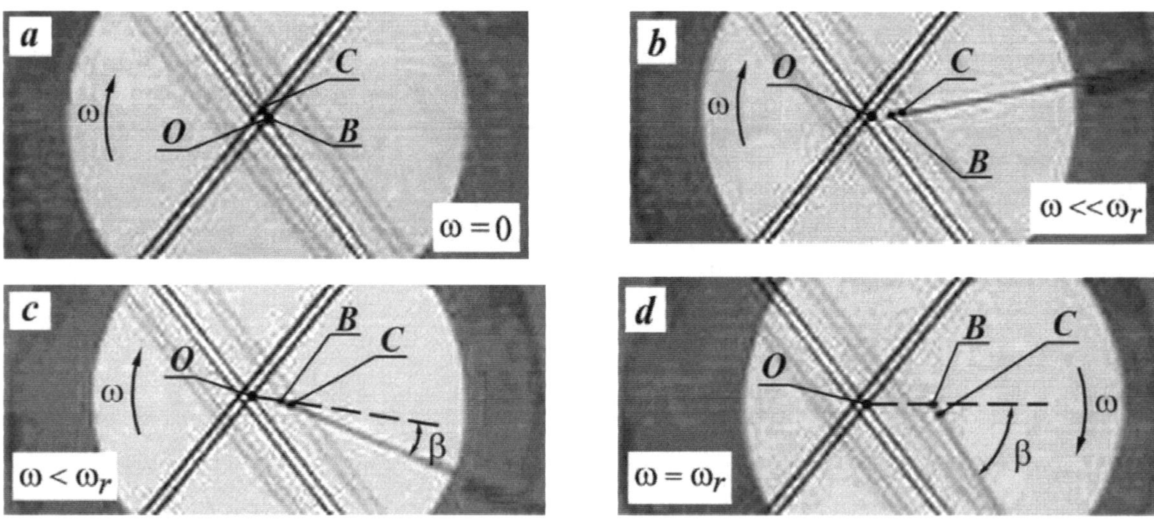

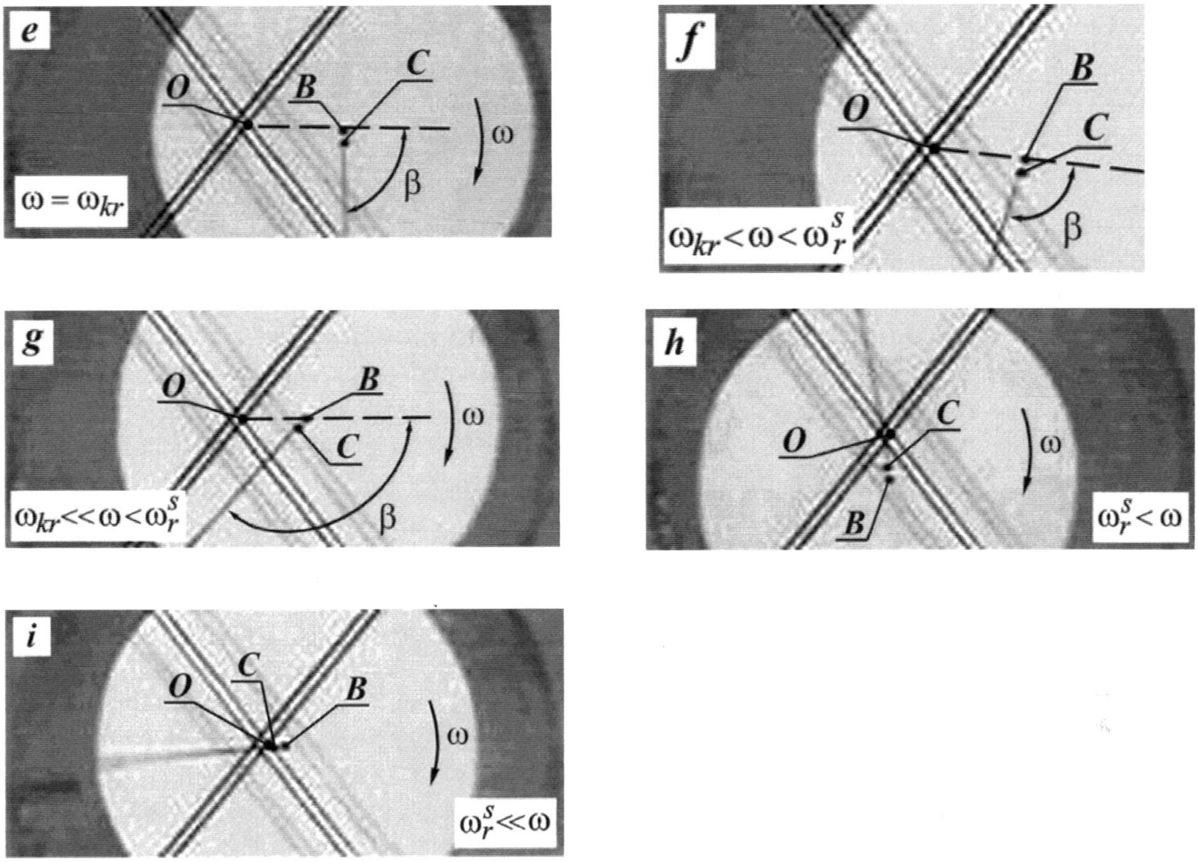

Fig 8

Conclusion

1. Study has shown that **1st critical speed** of the system's with flexible supports **forced resonance frequency response, is not equivalent to** rotor fundamental harmonic **resonance frequency** of flexible rotor in a free state in gravity environment

2. The operating speed frequency, and the efficiency of an operating horizontal rotating machine consisting of "rigid" or "flexible" rotor, constrained by gravity in two bearings within a finite "L" distance between them, should not be higher than 0.5x of rotor natural resonance in order to avoid any external pulsating excitation force, which could destabilize rotor operating in oil bearings, and excite subsynchronous whirling.

3. For a stable rotor operation at rated speed and load, in rotating machine system, the total system stiffness and mass including rotor, bearing oil and bearing support, must have 1st critical resonance response at system critical velocity 10% to 20% lower, than 0.5 frequency wave of rotor natural resonance. At the same time a resultant radial deviation,

between journals axis and rotor body centroidal mass axis must be brought to " balance" (symmetry) on balancing machine, between centrifugal forces generated from static eccentric mass axis, and dynamic mass axis created from three correction masses, proportionally distributed axially on rotor body between constraints.

4 When rotor is accelerated above supercritical velocity frequency, mass axis (non-rotating reference frame is a SCL- precession up to 90 degrees and gravity stability axis at 180 degrees. Dynamic orbit centerline (non-rotating frame - inertial stability axis through supercritical angular velocities and machine load increase at constant velocity) is **continuously self-centering*** in space, forced by inertia forces from the residual axially distributed mass eccentricities. Journals centers during mass axis self-centering at same time are whirling as gyroscopic rotating pendulums, pinned at COM and flexibly constrained at respective bearings and supports, proportional to torque moments at particular rotor angular velocity.

5. The objective of balancing rotors is to vanish the bearings dynamic reaction forces, and reduce journals radius of orbital motion at operating speed to magnitude of internal, inherent mass axis eccentricity. Accordingly, balancing methods of large turbine and generator rotors should be balanced, not based on axially distributed point mass "unbalance vectors", but based on compensating the inherent longitudinal mass axis, between bearings constraints with rotor in state of rest, by dynamic forces of compensating masses axially distributed in three planes, forming a dynamic mass axis mirror imaging the internal, inherent, mass axis (line intersecting centers of masses of rotor Null mode harmonic.

References

[I] Timoshenko S.P., Oscillations in Engineering, Moscow: State Publishing House of Physical ad Mathematical Literature, (1959) pp. 439.

[2] Dimentberg F.M., Shatalov K.E., Gusarov A.A., Oscillations of Machines, Moscow: Mashinostroyeniye, (1964) pp.308.

[3] Kozheshnik J., Dynamics of Machines, Moscow: State Publishing House of Machine-Building Literature, (1961) pp. 424.

[4] A.Y. Zhivotov. New Theory of Rotor Dynamic s: Dynamics of thick Rotor with Static Unbalance // IFToMM Sixth International Conference on Rotor Dynamics. - Sidney, Australia. - Proceedings. - 2002. - p. 1057.

[5] Zhivotov A. Y. New theory of rotor dynamics: Dynamics of Jeffcott Rotor with Moment Unbalance//12th IFToMM World Congress. - Besançon, FRANCE. - Proceedings - Volume I. 2007

[6] Ehrich, Fredric F. (1993): Balancing Method and Product; General Electric Company,

assignee. Patent US 5214585 A. 25 May 1993

[7] Schneider, Hatto. (2006): "Balancing and Beyond", Schenk Ro-Tec Gmbh.

[8] Kellenberger, W., "Should a Flexible Rotor be Balanced in N or (N+2) Planes?"; Trans. ASME, Engineering For Industry., 94, No 2 (1972).

[9] Abeyaratne, Rohan, 1952 –Lecture Notes on The Mechanics of elastic solids. Volume II: Continuum Mechanics/ Rohan Abeyaratne – 1st edition – Cambridge, MA and Singapore: ISBN-13:978-09791865-0-9

[10] Robert G. Loewy, Vincent J. Piarulli, "Dynamics of Rotating Shafts"; The Shock and Vibration Information Center, Naval Research Laboratory, Washington , DC, 1969

[11] J. Woodhouse, J. C. Rene, C. S. Hall, L. T. W. Smith, F. H. King, and J. W. McClenahan; Research Article: "The Dynamics of a Ringing Church Bell" ; Department of Engineering, University of Cambridge, Trumpington Street, Cambridge CB2 1PZ, UK

[12] Dwight E. Neuenschwander:" Emmy Noether Wonderful Theorem, 2011 The John Hopkins University Press

[13] Ludwig von Bertalanffy, "Systems theory"; https://en.wikipedia.org/wiki/Systems_theory

[14] Stephen H. Crandall; Chapter 1; "Nonlinear Dynamics and Stochastic Mechanics" edited by Wolfgang Kliemann and N. Sri Namachchivaya

[15] Raj Subbiah, Jeremy Eli Littleton; Rotor and structural dynamics of Turbomachinery, Springer

[16] Racic Z, Racic M. (2014): "Development of a New Balancing Approach for Significantly Eccentric, Bowed Rotors" In: Mechan. Machine Science, Vol. 21, Paolo Pennacchi (Ed);

[17] 8. Racic Z, Racic M. (2014): "Behavior of Eccentric Rotors through the Critical Speed Range." In: Mechan. Machine Science, Vol. 21, Paolo Pennacchi(Ed): Proceedings of the 9th IFToMM International Conference on Rotor Dynamics, Milan, Italy, 22-25 Sept 2014.

COMBINING WEAR DEBRIS AND VIBRATION FOR A MORE COMPLETE UNDERSTANDING OF MACHINERY HEALTH

Jeremy Sheldon, jeremy.sheldon@poseidonsys.com
Ryan Brewer, ryan.brewer@poseidonsys.com
Stephen Steen, stephen.steen@poseidonsys.com
Poseidon Systems, LLC, 200 Canal View Blvd, Rochester, NY

Abstract: This paper thoroughly discusses vibration and oil debris monitoring. The application of each sensing technology to an actual real-world wind turbine application will be presented in detail. A focus of the paper will be the use of vibration and oil debris together to improve overall CBM metrics. By combining oil debris and vibration, the CBM practitioner will realize improved visibility to the health of their target machine.

Keywords: Oil debris, Vibration, Wear, Condition Monitoring

Introduction:

Financial, environmental, and resource pressures faced by most industries are causing increased interest in improved reliability centered, and by extension predictive, maintenance practices. Many of the sensors, methods, and techniques used to realize the benefits of reliability centered maintenance have been available for many years. The advent of the internet of things era has helped reduce some of the technological and cost barriers associated with implementing these techniques. In fact, in the IOT era, there has been an explosion of new technologies and applications of those technologies for enabling predictive maintenance, especially those focused on machinery condition monitoring.

Condition monitoring uses various sensors and analysis methods to detect or infer the health of machinery. Temperature, current monitoring, vibration, oil analysis and sensing, loads monitoring, and many other sensing methods can be used to measure parameters correlated to machinery health and machinery failure modes. Machinery OEMs, owners, and operators faced with the pressures are tasked with finding ways to mitigate the risk associated with inoperable machinery. Because of this, OEMs, operators, and institutions have been investing in these sensing technologies to detect faults and prevent them from progressing into expensive failures [5]. The goal of condition monitoring systems is to detect faults as early as possible in their failure timeline to enable the lowest cost repair possible.

Each sensing technology and its interpretation offer varying performance regarding fault detection and isolation. As shown notionally in Figure 1, various sensing technologies offer various detection horizons. Those implementing a condition monitoring system (CMS) must evaluate the sensors, fault detection performance, cost, and various other aspects before selecting a CMS. Wear debris and vibration are typically the first indications of a developing fault and have the potential to allow for addressing machinery faults earlier. In wind turbine gearboxes, one of the most valuable actions to detect a fault early enough to allow low-cost up-tower solutions and avoid a full gearbox replacement. Examples of both vibration and oil debris monitoring are presented below. Following the individual examples is a combined real-world case study of how combining oil debris and vibration can offer a more complete CBM solution.

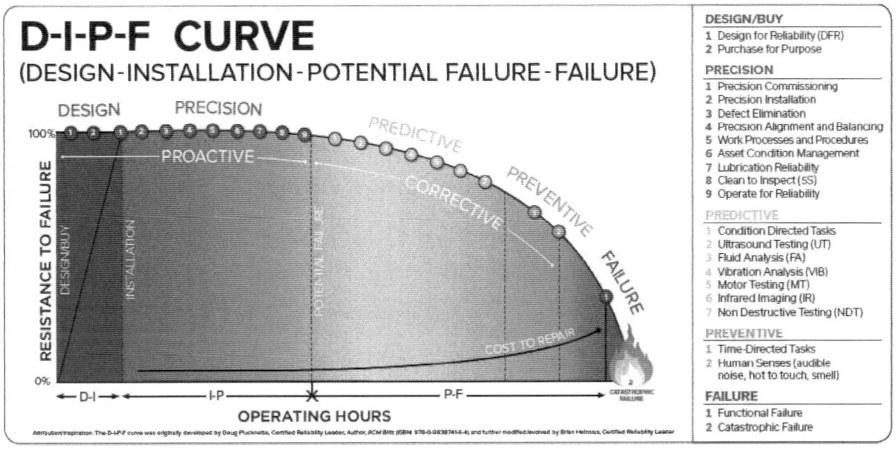

Figure 1 Progression of Fault and Detection Methodologies [1]

Vibration Overview:

Vibration analysis has proven to be one of the most reliable and quantitative indicators of gearbox component failures. Acquiring, analyzing, and monitoring of a machinery's vibration can reveal much about that machine's health. Typically, vibration is measured via an accelerometer type sensor mounted directly to the machine. The measured raw vibration signal must first be analyzed in order to extract useful information form the acquired waveform [6,7]. Many analysis techniques exist, usually applying one or more signal processing techniques to process the time waveform into useable reduced-order information, sometimes called features or condition indicators. As part of this analysis, shaft speeds and machine geometry are often used to calculate component-specific frequencies useful in fault isolation. This paper will not get into the finer details of vibration analysis but will instead focus on the application and results from real world examples.

A commonly used vibration indicator, or feature, is root mean squared (RMS). RMS is a relatively easy to compute statistical feature that is indicative of the overall broad band vibration level. One of the limitations of RMS is its lack of fault isolation capabilities when compared to methods involving frequency analysis, such as FFT analysis, which will be described later in the paper. Nonetheless, RMS is an important part of many machinery monitoring systems, including integral to accepted standards, such as ISO-10816-21 [2].

As part of a capability assessment, the authors installed two MEMS-based wireless vibration nodes, hereafter referred to as vibration nodes, on each of three different 1.5 MW wind turbine generators. The sensors were installed radially orientated near the drive (nearest the main rotor) and non-drive end (furthest from the main rotor) generator bearings. The focus of the installation was to assess the capabilities of the sensors and integrated system. As such, the sensors were installed on three generators of known health states: healthy bearing, incipient bearing fault, and severe bearing fault. The faulted bearings were earlier identified and confirmed via borescope inspection. Data from the vibration nodes was collected over approximately eight months to assess the vibration system's performance on each generator, mainly focused on the bearings. The monitoring period extended beyond the bearing replacement to compare before and after repair vibration levels.

The three RMS trends over time are shown in Figure 2. As expected, higher vibration levels were observed on faulted generator bearings during the first few months. The generators with faulted bearings were inoperative while repairs were in progress, the healthy turbine continued operating. Once repairs were complete, all generators exhibited similar vibration levels, with the exception of a short duration of higher vibrations during the first start up. From the clear difference in the before and after repair vibration levels confirms the repair adequately remedied the underlying fault state. However, from the trend, it is difficult to tell the difference between the fault severities, which is important for understanding the likely fault progression and timeline in which repairs are required.

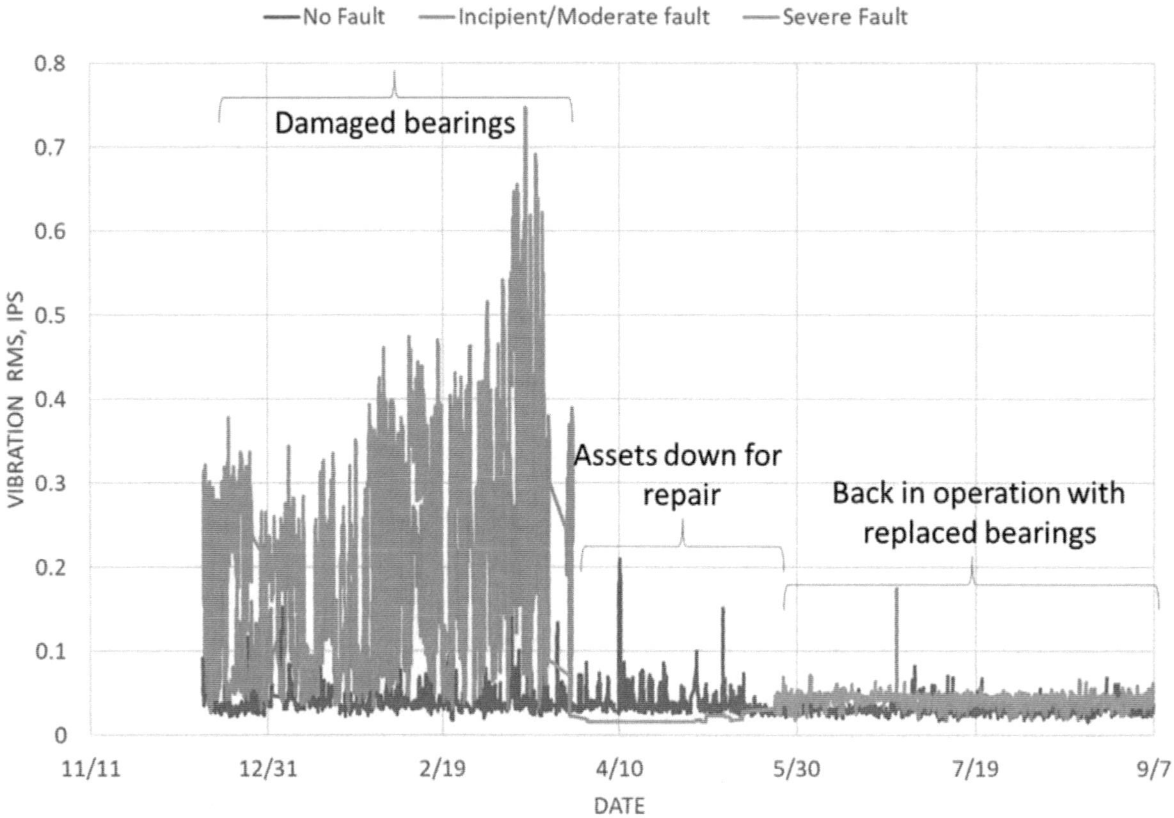

Figure 2 Generator Bearing Vibration Trend

An often-useful method for comparing vibration data from different assets is via a statistical box and whisker plot. These well-defined plots allow quick comparison of data from multiple assets to identify the anomalous assets and assess likely relative fault severity. Figure 3 shows the data from the three generators across the eight months of monitoring. There are three primary conclusions that can be drawn from these plots. First, the generators with the incipient/moderate fault and severe fault have clearly elevated vibration levels compared to the healthy generator. Second, the incipient/moderate bearing fault is producing lower vibration than the severe bearing fault, thus correlating the vibration level to the known fault level. Third, both the severe and incipient/moderate bearings faults appear to be progressing during the first four months of monitoring, as suggested by the upward trends in Figure 3 and Figure 2.

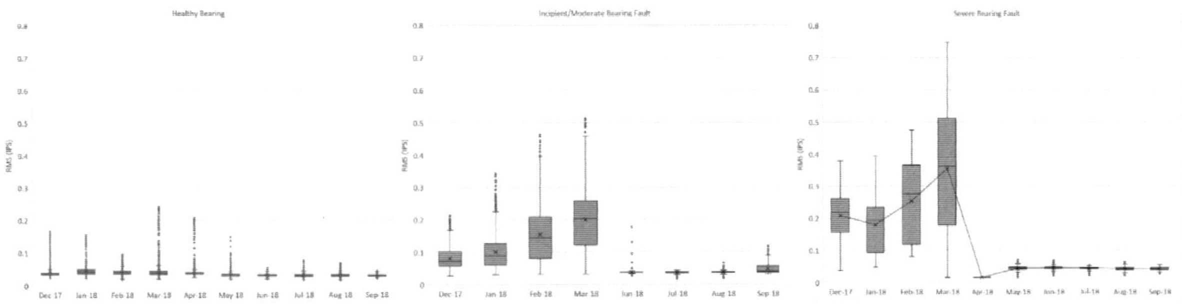

Figure 3 Generator Bearing Vibration Box Plots

From another example of vibration monitoring, the vibration nodes were installed on multiple wind turbine main bearings as part of a larger condition monitoring effort. The main bearing is a very large slowly rotating bearing that supports the main rotor of the wind turbine, ideal for monitoring using more basic vibration techniques. Approximately six months of data from four of the main bearings monitored are shown in Figure 4. Poseidon Systems noticed Tower Four (T4) was trending above the remaining assets in early February. The trend continued and over the next few months, T4 vibration levels continued to increase, as shown in Figure 5, at which point the site operator was contacted and they confirmed the bearing damage via borescope. Vibration monitoring then enabled the operator to safely operate with the faulted bearing until a convenient, cost-effective date for bearing replacement. Without the insight into the evolving condition of the bearing, continued operation would have been risky due to the potential for rapid degradation and potential bearing seizure.

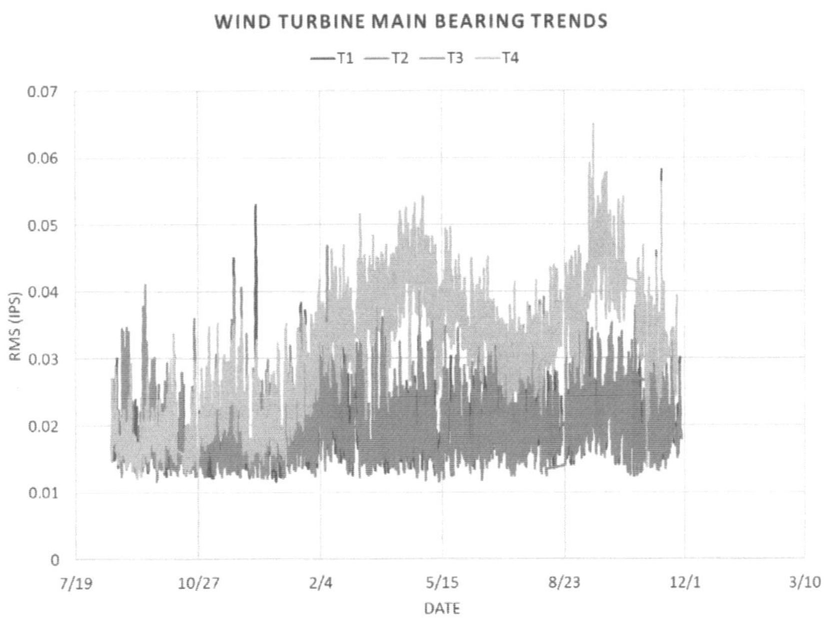

Figure 4 Main Bearing Vibration Trend

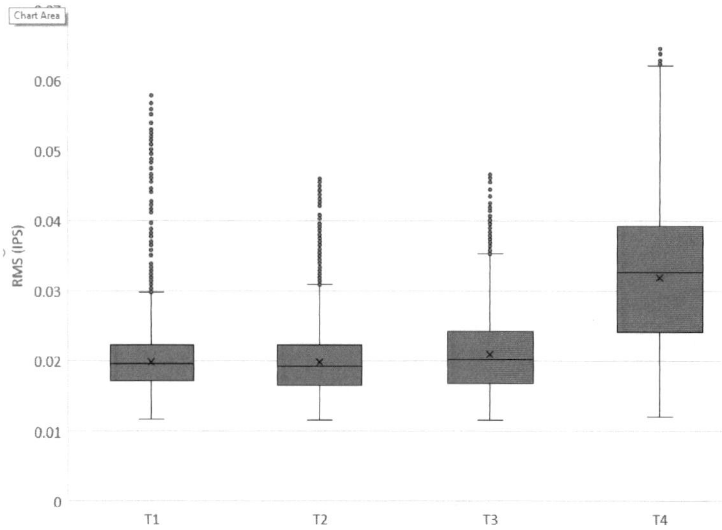

Figure 5 Main Bearing Box Plot Comparison

Oil Debris Overview:

Online wear debris sensors provide the unique capability of constantly monitoring the levels of ferrous and nonferrous wear debris particles within a mechanical system. Accelerating wear generation rates and momentary surges in particle counts can be easily recognized and acted upon before a critical gearbox failure occurs. Wear debris monitoring is a more direct indicator of component health than vibration and usually requires much less analysis. The utility of online wear debris monitoring as a means of detecting mechanical component failures and continually assessing their severity makes them an attractive alternative to vibration monitoring systems in many applications, especially within oil-lubricated gearboxes.

Various oil debris monitoring technologies are available for online measurements. Inductively Coupled Plasma Atomic Emission Spectroscopy (ICP-AES) or analytical ferrography methods are commonly used in laboratory settings [3]. Optical based particle counters are common in labs but can be found in certain field applications. Magnetic-based chip collectors and detectors are typically used in gearbox monitoring applications [4]. Another technology used for debris sensing is based on an inductive sensor created by one or more coil windings around a non-conductive fluid passage (sensing bore). The coil is excited with a high-frequency oscillating voltage and the resulting magnetic field couples with any metallic particle present within the field. The particle's passage results in a measurable impedance change in the sensing coils providing information about the particle's size, type (ferrous or nonferrous), and speed. Unlike many other sensing technologies, the output of wear debris is a direct measurement of mechanical damage. In addition, the results are easy to interpret with little expertise needed to analyze and understand.

Mechanical faults can propagate quickly, developing from a state of low risk of major component failure to high risk in a matter of days, if not hours. Online metallic debris monitoring enables these changes in severity to be observed and tracked in real-time, allowing adjustments in operation to prevent catastrophic failure and extend operating life until a repair or replacement can occur. Furthermore, wear debris data can be correlated with machine usage, temperature, vibration, and other sensor data for further insight into the health of the machine. For example, a large wear debris increase could correspond to a severe load event, perhaps due to high winds in the case of a wind turbine gearbox or jammed crusher in a mining application. It is useful to summarize oil debris sensor outputs in two key life gauge measurements rate of material loss (mass per unit time/flow, sometimes called concentration), and fault severity (cumulative mass). An example from a failing wind turbine gearbox is shown Figure 6. The rate of material loss is sporadic as wear material generation rates change with gearbox loads and speeds, while the cumulative mass steadily increases over time. Larger peaks in the concentration level are indicative of more severe wear events, and increasing slope of the cumulative mass also indicates increase fault severity.

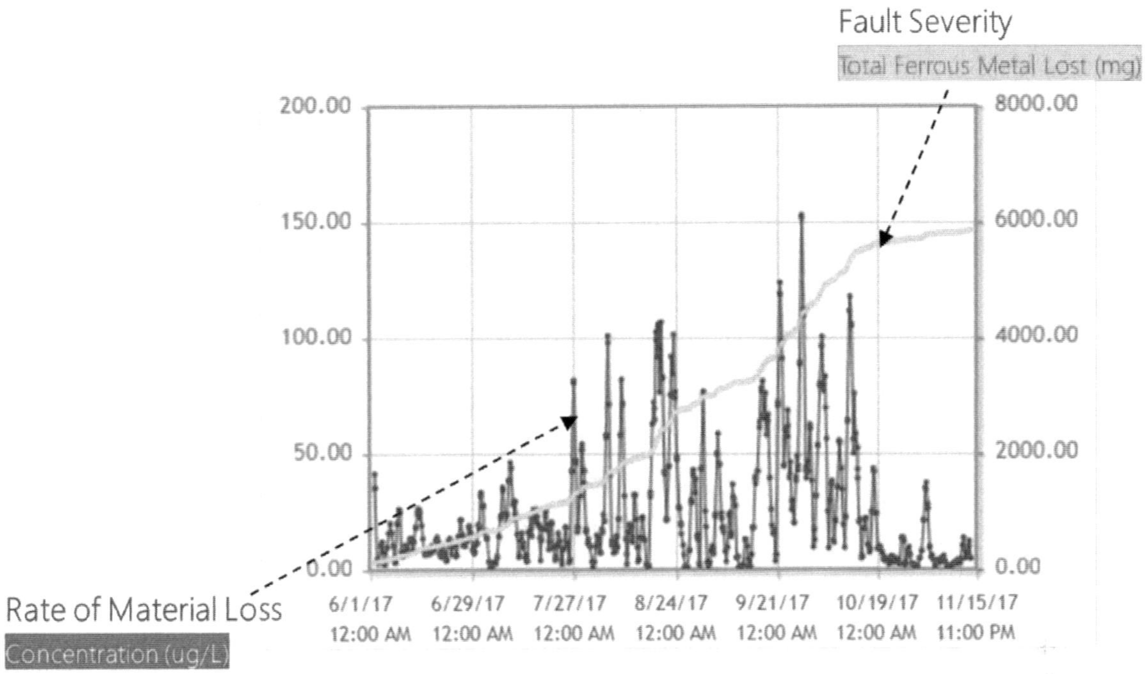

Figure 6 Example Oil Debris Trend

An inductive-based oil debris monitor was installed on several large industrial gearboxes at a customer site. Although several gearboxes were monitored, only one is shown herein for brevity. The ferrous wear debris concentration for a 24-hour interval data from several months is shown in Figure 7. The concentration levels steadily increase from February through June at which point they reach warning levels. The alarm and warning levels were statistically derived from a large data set that including healthy and known faulted gearboxes. Notice the fluctuations of the concentration values are due to the variable speed and loading of the gearbox.

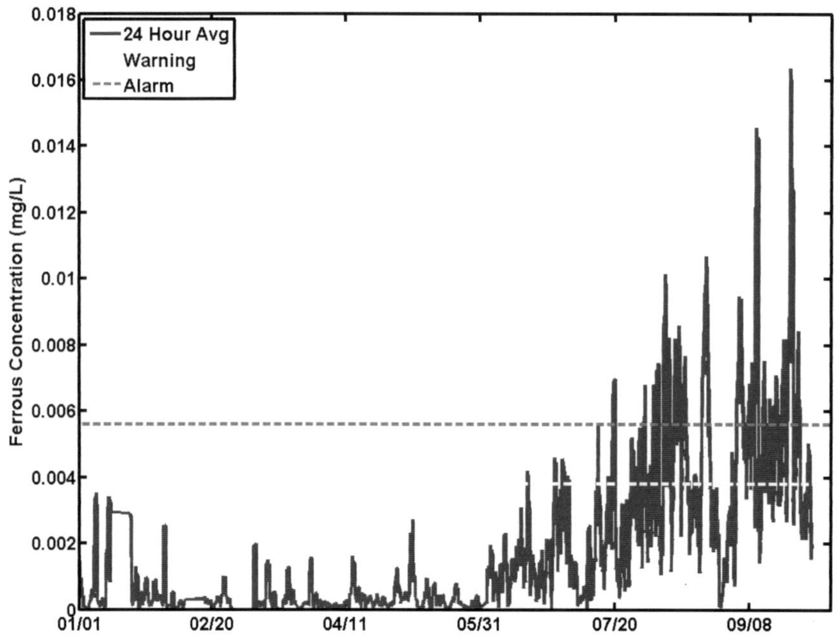

Figure 7 Industrial Gearbox Wear Concentration Trend

The debris sensor installed had the advantageous capability to count and size the sensed debris in a sequence of size ranges. Understanding the sizes of the particles, both in counts and size distribution, provides valuable information that can be used to characterize fault type, fault severity, and even fault location in many instances. Figure 8 shows the number of particles observed in each month in four key size ranges. During fault initiation (February through May), the wear generation is almost exclusively sub-200µm particles. As fault propagation increases, (June through September), the distribution shifts to include larger particles, with an increase in the smaller size debris as well. These distributions and their evolution over time are important fault signatures that can be used to diagnose fault severity and location, for example, a gear fault signature looks different than a bearing fault signature. Based on the trend and the distribution the operator was notified of a progressing fault.

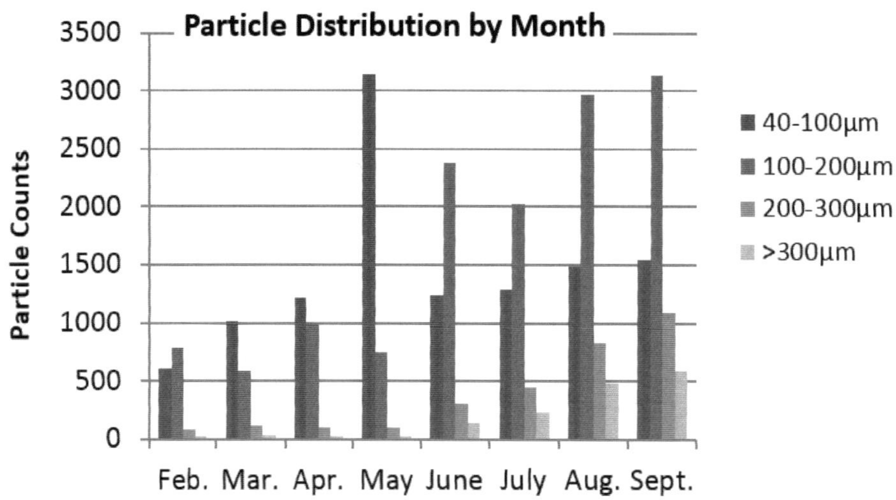

Figure 8 Industrial Gearbox Wear Size Distribution

The resulting borescope inspection revealed a fractured tooth on the low-speed helical pinion, as shown in Figure 9. Unlike many detected fractured teeth which can lead to very fast catastrophic failure, this gear fault was relatively benign. The maintenance team was able to perform an in-situ repair to prolong the life of the gear. Fault identification occurred early enough to prevent secondary damage to the mating bull gear, significantly reducing the overall cost of repair. Post repair debris data was used to confirm the repair and monitor the gearbox for the remainder of its life ensuring safe operation.

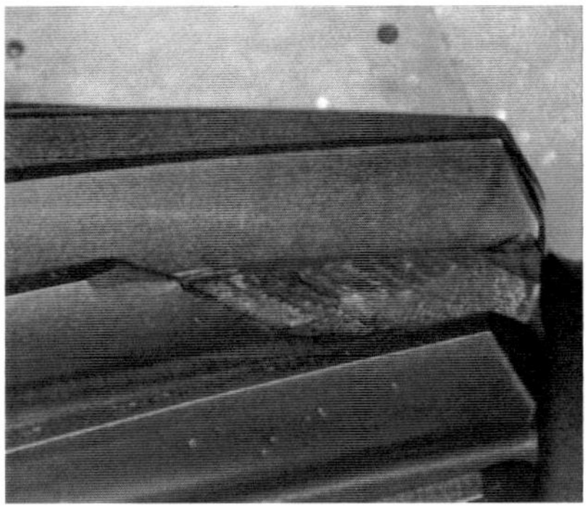

Figure 9 Industrial Gearbox Damage

Field Case Study:

The above vibration and oil debris examples highlighted the benefits of the technologies individually. Combining the strengths and weaknesses of both provides a more complete condition monitoring approach. The following case study will show these combined benefits in the context of an actual in-field fault event. An oil debris sensor and vibration monitoring system were installed on a 1.5 MW wind turbine gearbox. The oil debris sensor was installed many years before the other instrumentation was added. Early in the timeline, a borescope inspection was conducted at recommendation based on the debris monitoring and noted damage to the ring gear, planet bearings, high-speed shaft bearings, and high-speed gear. Based on the borescope and long-term oil debris monitoring data, the operator scheduled a gearbox rebuild. Until the repair, the wind turbine remained in operation with the various sensors installed to monitor for impending catastrophic failure over the remaining years and months of operation.

An oil debris sensor was installed in approximately April 2016. Data from the sensor is measured every 10 minutes to provide near real-time wear debris assessment. Within the first few months following installation, abnormal amounts of debris were measured and the 24-hr concentration crossed the established limit, as shown in Figure 10. A borescope inspection was conducted in February 2017 to confirm the damage. The borescope report noted that the ring gear had widespread abrasive wear and macro pitting on the teeth, the planetary bearing had heavy macro pitting, the HSS upwind bearing had abrasion and score marks due to debris in the oil, and there were markings on the high-speed pinion, also due to debris in the oil. The gearbox continued to operate for more than 2 years until it was repaired in July and August of 2019. The debris sensor was used during this time to continuously monitor the wear generation rate and distribution to ensure the faults did not propagate too rapidly and lead to catastrophic failure.

Once repaired the gearbox was put back into full service. As expected, the debris generation greatly reduced after repair. A zoomed-in plot showing only data near the gearbox repair is shown in Figure 11. By monitoring debris generation both before and after the repair, condition monitoring can confirm repair efficacy as evidenced by the clear reduction in wear debris measured post-repair. Notice a relatively large spike in debris measured right at start up, this is typical as the sensor measures the debris left in the gearbox from new components and the rebuild process.

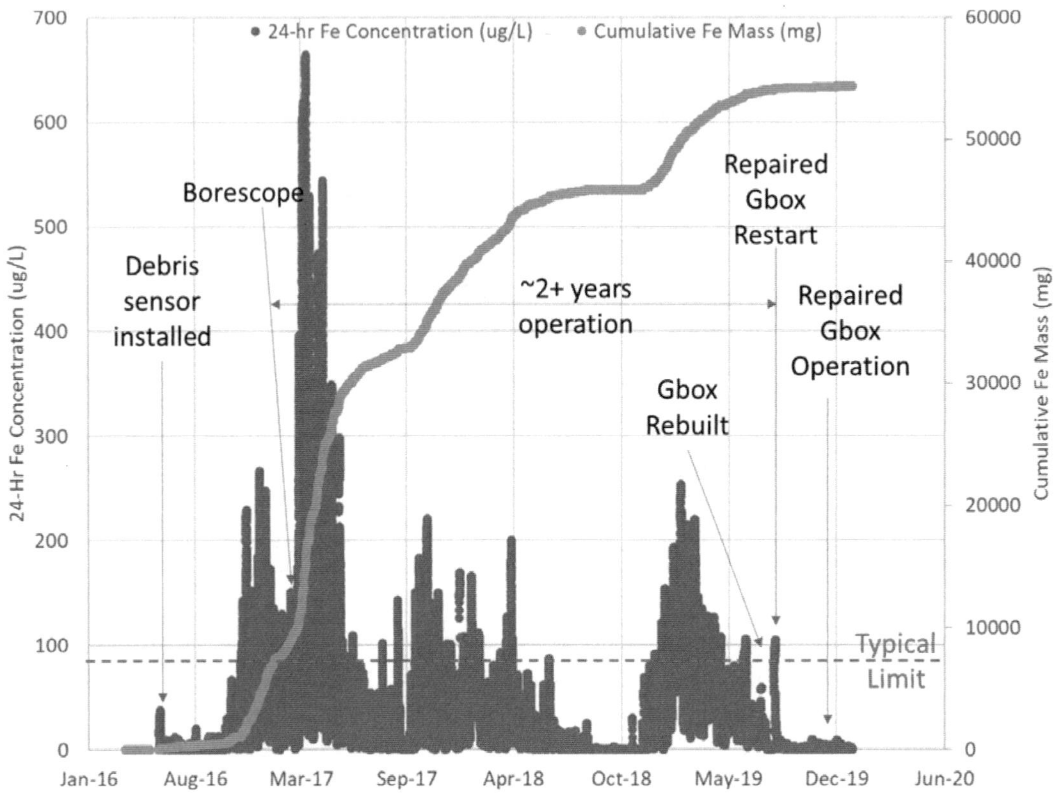

Figure 10 Field Case Oil Debris Trend, All Time

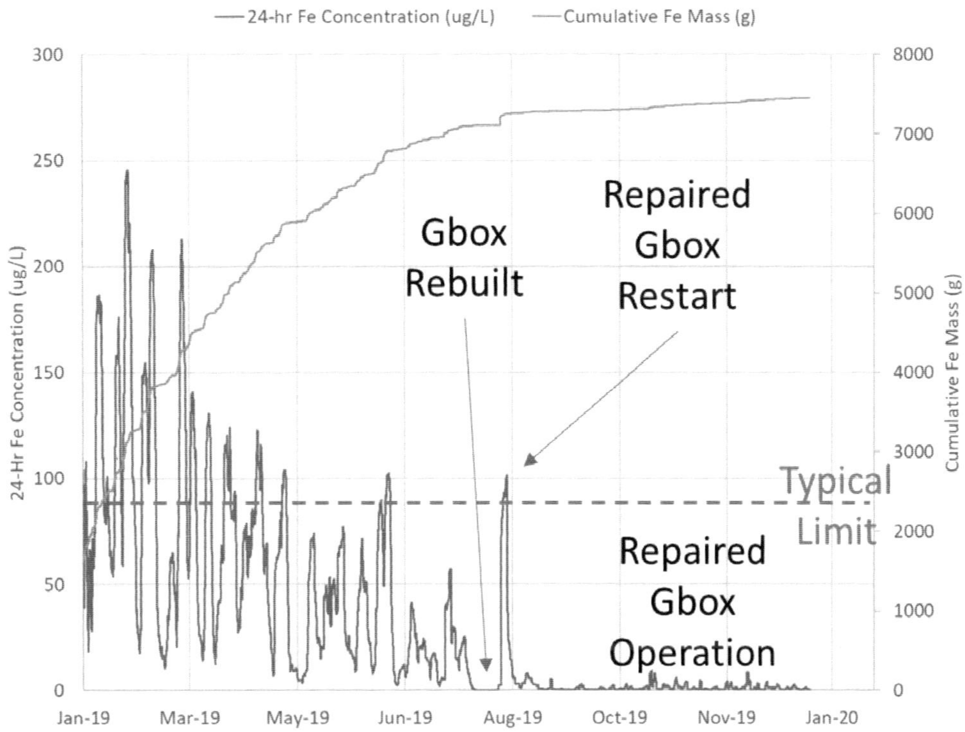

Figure 11 Field Case Oil Debris Trend, Months Prior and After Repair

Oil debris sensors were supplemented with vibration monitoring for the last few months of the gearbox life and the first few months following the repair. Multiple vibration nodes were also installed on the gearbox for the three months prior to replacement to provide additional insight into the final stages of the gearbox failure. These nodes monitored the gearbox periodically, mainly providing RMS measurements of the vibration once an hour. The RMS measurements are overlaid with the oil debris data in Figure 12. Overall, the vibration was reduced after the repair, however, the pre-post repair difference in vibration levels compared to the difference in debris measured is notable. There was a much larger decrease in the amount of debris measured than there was a decrease in vibration upon restart. The oil debris monitor appears to be more sensitive to the faults than the RMS measurement, at least in this application. In addition, one can directly see that the vibration is generally correlated to the debris concentration. During times of high vibration, there is more wear debris generated and measured, especially in the damaged gearbox. A more rigorous correlation of the vibration levels to the debris could be a useful metric for determining changing health state.

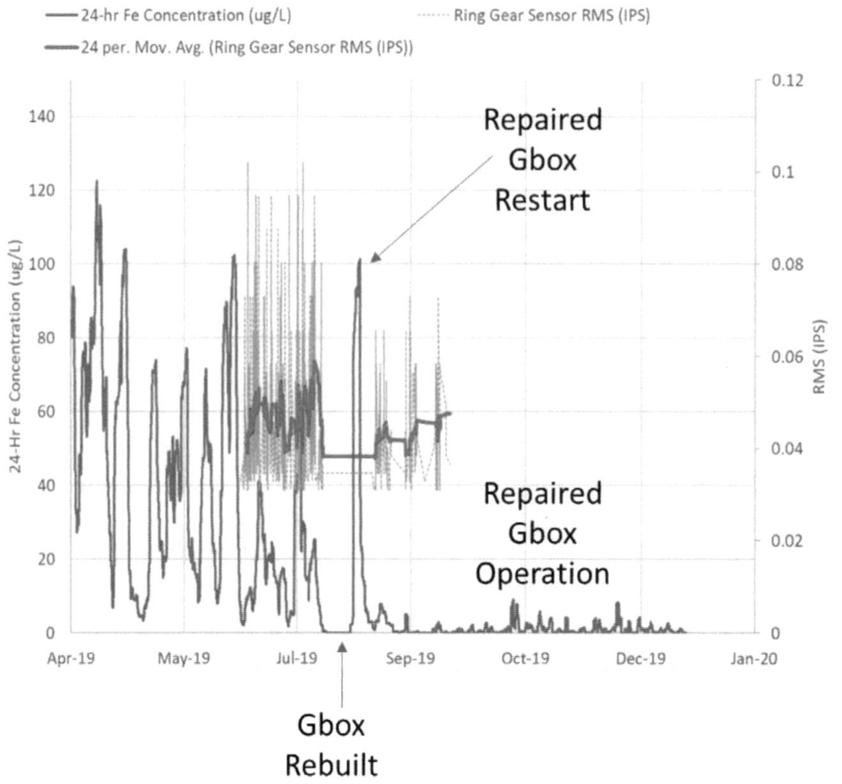

Figure 12 Field Case Oil Debris Compared to Vibration

Spectral and time domain analysis of the data from the vibration nodes was also performed. A sample fast Fourier transform (FFT) from all three vibration axes before the repair is shown in Figure 13. Although the intermediate speed (IMS), high speed (HSS), and ring gear mesh frequency (GMF) were all visible, the FFT did not show clear indications of the faults. No sidebands or multiple harmonics were clearly visible as one would expect from a faulted gear. However, this is not all that surprising given the MEMS-based vibration nodes relatively low bandwidth and that there was no additional processing applied to the data. Thus, indicating these as configured nodes may not be ideal for gearbox fault isolation. Although less than ideal, there were indications in the time domain data of a bearing and/or possibly a gear issue The impact type of vibration events in Figure 14 are consistent with a bearing fault, such as spalling, but further analysis of the vibration node data did not provide additional fault isolation.

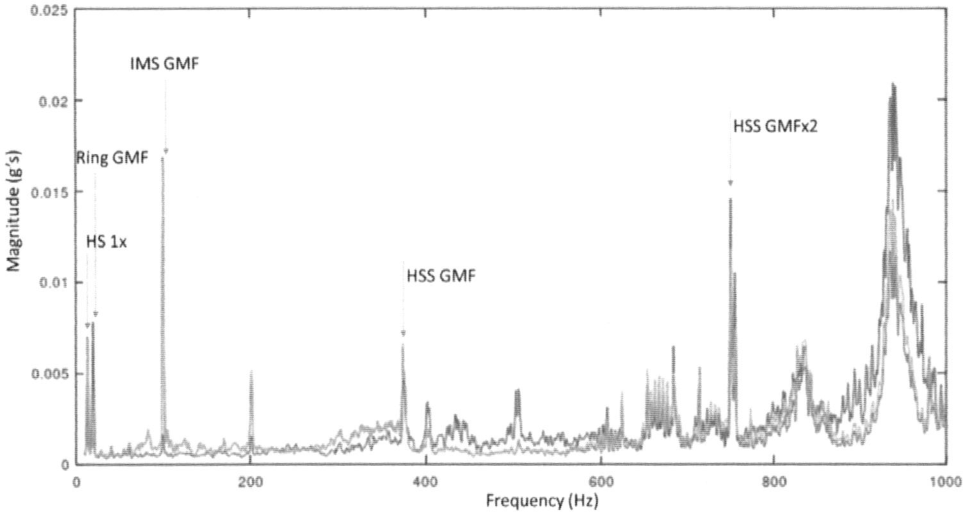

Figure 13 Field Case Oil Faulted Gearbox FFT Examples

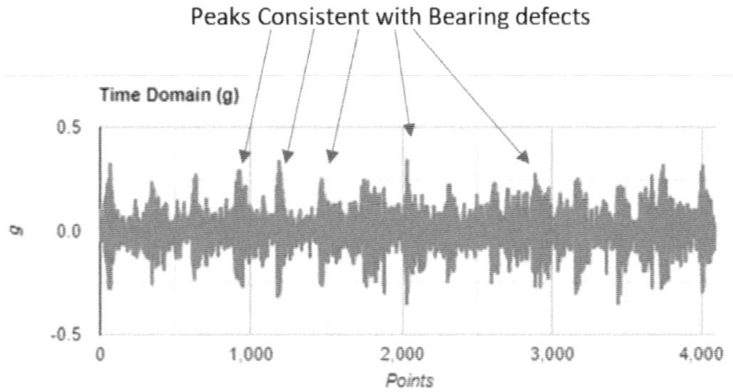

Figure 14 Field Case Faulted Gearbox Time Series Examples

To address the fault isolation component, a high-performance vibration monitoring system was also installed on this gearbox at the same as the vibration nodes to further isolate the known faults. This system has sensors with high bandwidth and integrates advanced signal processing and feature extraction techniques, including time-synchronous averaging for gears and enveloping for bearings [8]. Results of the more advanced analysis showed clear signs of many different faults, including intermediate gear, high-speed gear, and some bearing issues, confirming what the debris and vibration nodes were indicating. For example, the enveloped spectrum in Figure 15 clearly shows abnormal high-speed rotor side (HSS) bearing defect frequencies magnitudes; one would not expect to see such defined peaks unless a bearing fault exists. Intermittently high amplitude modulated (AM) Kurtosis, such as Figure 16, also confirmed the presence of gear faults.

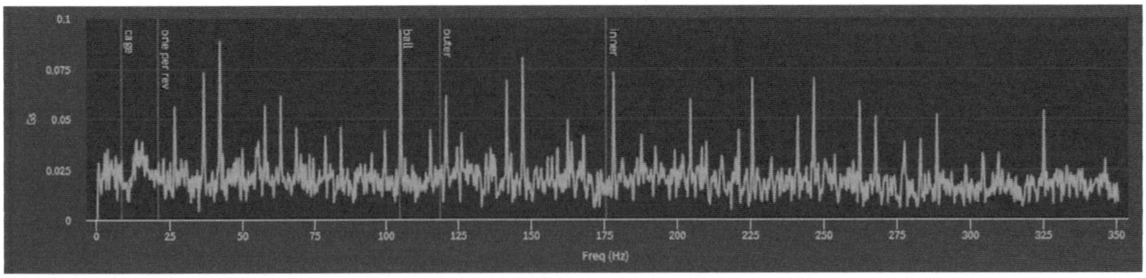

Figure 15 Field Case, Enveloped Spectrum, HSS Rotor Side Bearing Defect Frequencies

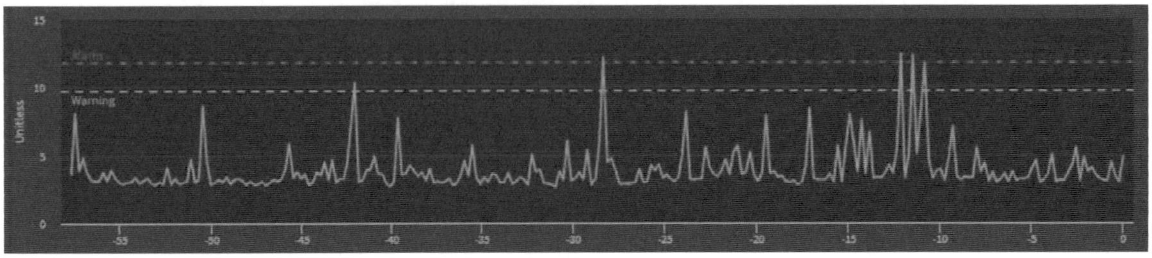

Figure 16 Field Case, Intermediate Pinion Gear Feature

As described above, the targeted gearbox was repaired and returned to service. After returning to service all sensors indicated an improved health state. To a varying degree, each of the sensor systems indicated the gearbox was faulty and showed a return to normal levels after repair. Generally, the oil debris data provided the clearest and easiest to directly interpret indications of the gearbox health state. The vibration analysis was useful in further isolating the faults and when combined with the oil debris, can be used to monitor fault progression based on the wear debris measured. Due to the short duration of this install, there was not an opportunity to monitor much fault progression while the vibration systems were installed.

Conclusion

The challenges surrounding detecting and predicting faults in complex systems can be an issue since this is one of the key purposes of condition monitoring. This case study illustrates how a combination of oil debris analysis and vibration analysis serves as the best way to serve this key function. As shown, oil debris analysis is able to provide indications of abnormal health which can allow the operator to have more time to plan an optimal plan for gearbox replacement. The vibration analysis enabled precise tracking of component-specific fault indicators to help monitor the rapid progression of faults. As shown above, when these methodologies are combined, it creates an environment where faults can be detected and predicted with much more information that can be provided by other sensing methods to give a more complete understanding of the evolving machinery health state.

There is no single sensor type or analysis method that will work for every component and every failure mode. The technique of combining techniques and sensors applied in this paper will likely result in the best results for any condition monitoring application. One must always balance cost and complexity with the CBM metrics like detection horizon, false alarms, etc. In addition, certain components and machines will require specific monitoring. For instance, oil debris cannot be used on grease-lubricated components, and vibration systems cannot detect clocking bearing failure modes. Unfortunately, budgets often restrict what is possible, often requiring selection of only one monitoring approach. In that case, one must be careful in selecting the approach as there is no single sensing solution capable of detecting every failure mode.

In order to get a complete understanding of machine health, it is important to have a large amount of high-quality data. Combining multiple sensing techniques is the best way to do this as they each provide their own value. This is true for any industry, however oil debris monitoring and vibration analysis are an excellent example to show the value, as has been described in this paper through the case studies and empirical data.

Acknowledgments

The authors would like to acknowledge our customer site for supporting the data collection and Eric Bechhoefer for providing expert knowledge and assistance in analyzing this data.

References

1. Reliabilityweb.com. Copyright 2016-2018. NetexpressUSA Inc. d/b/a Reliabilityweb.com, accessed January 9, 2020.
2. ISO 10816-21, "Mechanical vibration — Evaluation of machine vibration by measurements on non-rotating parts — Part 21: Horizontal axis wind turbines with gearbox," First edition, May 1, 2015.
3. Booser, E. Richard, "Tribology Data Handbook: An Excellent Friction, Lubrication, and Wear Resource," CRC Press, Boca Raton, FL 1997, pg 900-907.
4. "Engineered Sensors Tedeco Products Lubrication System Diagnostics Sensors Magnetic Chip Collectors," http://www.tedecoindustrial.com/mag.htm.
5. Coronado, Diego, Fischer, Katharina, "Condition Monitoring Of Wind Turbines: State of the Art, User Experience and Recommendations," Fraunhofer Institute for Wind Energy and Energy System Technology IWES, January 2015.
6. Sheldon, Jeremy S., Watson, Matthew J., Byington, Carl S., "Integrating Oil Health And Vibration Diagnostics For Reliable Wind Turbine Health Predictions," Proceedings of ASME Turbo Expo 2011 GT2011 June 6-10, 2011, Vancouver, British Columbia, Canada.
7. Sheng S., H. Link, W. LaCava, J. van Dam, B. McNiff, P. Veers, and J. Keller, S. Butterfield and F. Oyague, "Wind Turbine Drivetrain Condition Monitoring During GRC Phase 1 and Phase 2 Testing," National Renewable Energy Laboratory, 2011.
8. Bechhoefer, E., Kingsley, M., "A Review of Time Synchronous Average Algorithms," Annual Conference of the Prognostics and Health Management Society, 2009.